BURLEIGH DODDS SCIENCE: INSTANT INSIGHTS

NUMBER 36

Reducing antibiotic use in dairy production

I0130695

burleigh dodds
SCIENCE PUBLISHING

Published by Burleigh Dodds Science Publishing Limited
82 High Street, Sawston, Cambridge CB22 3HJ, UK
www.bdspublishing.com

Burleigh Dodds Science Publishing, 1518 Walnut Street, Suite 900, Philadelphia, PA 19102-3406, USA

First published 2021 by Burleigh Dodds Science Publishing Limited
© Burleigh Dodds Science Publishing, 2021. All rights reserved.

British Library Cataloguing in Publication Data
A catalogue record for this book is available from the British Library

ISBN 978-1-80146-165-8 (Print)
ISBN 978-1-80146-166-5 (ePub)

DOI 10.19103/9781801461665

Typeset by Deanta Global Publishing Services, Dublin, Ireland

Contents

Series list

Title	Series number
Sweetpotato	01
Fusarium in cereals	02
Vertical farming in horticulture	03
Nutraceuticals in fruit and vegetables	04
Climate change, insect pests and invasive species	05
Metabolic disorders in dairy cattle	06
Mastitis in dairy cattle	07
Heat stress in dairy cattle	08
African swine fever	09
Pesticide residues in agriculture	10
Fruit losses and waste	11
Improving crop nutrient use efficiency	12
Antibiotics in poultry production	13
Bone health in poultry	14
Feather-pecking in poultry	15
Environmental impact of livestock production	16
Pre- and probiotics in pig nutrition	17
Improving piglet welfare	18
Crop biofortification	19
Crop rotations	20
Cover crops	21
Plant growth-promoting rhizobacteria	22
Arbuscular mycorrhizal fungi	23
Nematode pests in agriculture	24
Drought-resistant crops	25
Advances in crop disease detection and decision support systems	26
Mycotoxin detection and control	27
Mite pests in agriculture	28
Supporting cereal production in sub-Saharan Africa	29
Lameness in dairy cattle	30
Infertility/reproductive disorders in dairy cattle	31
Antibiotics in pig production	32
Integrated crop–livestock systems	33
Genetic modification of crops	34

Chapter 1

Responsible and sustainable use of medicines in dairy herd health

David C. Barrett, Kristen K. Reyher, Andrea Turner and David A. Tisdall, University of Bristol, UK

1 Introduction

As an agricultural community we have a moral obligation to raise healthy farmed livestock, both to safeguard their welfare and to optimise the quality and safety of the food they produce. Veterinarians and farmers are not simply responsible for animal health and welfare, but also for public health, and can address both by acting to protect the food chain and the environment in which we live and work. These challenges are compounded by the need to produce an increasing amount of food to supply the growing global population (FAO, 2017), with the majority of the growth occurring in developing countries where newly affluent consumers seek greater quantities of milk, meat and eggs (Kearney, 2010). Elsewhere, many of the global poor are undernourished (FAO, 2015), and there is a need to improve nutrition by eating modest amounts of animal protein which can significantly improve child growth and cognitive development (Neumann et al., 2007). The nutritional, socio-economic and cultural benefits of ruminant protein production in sustainable agricultural systems are well recognised (CAST, 2013), as is the need to optimise resource use (e.g. land, water and fertilisers) while mitigating environmental impacts, including (but not limited to) soil loss, greenhouse gas emissions and nutrient run-off. These challenges must be met within the constraints of an increasingly competitive

http://dx.doi.org/10.19103/AS.2016.0006.26

global market, with a significant disconnect between urbanised consumers and agriculture, and where reliance on the Internet and social media sites for news and information allow food production myths to abound (Capper and Yancey, 2015).

The acquisition of milk requires harvesting an animal product on a daily basis, sometimes three or more times a day. In its natural state, milk has a very short shelf life, so it is thus processed into a myriad of products (e.g. cream, butter, yogurt, cheese, etc.), both to extend its shelf life and to diversify the products available to the consumer. The need for these products to be consistent in quality and to maximise their shelf life means that tight controls are placed on the levels of allowable bacterial contamination and somatic cell count (SCC) that are deemed acceptable in milk destined for human consumption. In order to achieve these targets – particularly SCC levels – both clinical and subclinical mastitis (often caused by bacterial intramammary infections) must be kept under control. This has historically meant the use of antimicrobials (AMs), both to treat clinical cases of mastitis during lactation and to treat existing and prevent new infections at the end of lactation, usually termed 'dry cow therapy'. Whilst this is still normal practice today in much of the world, for reasons which will be discussed later in this chapter, established practices are now being challenged, and will need to change in the future.

Furthermore, medicines – particularly AMs – must be prevented from entering the food chain, both because they may be directly detrimental to human health but also because they can interfere with bacterial cultures used in the production of dairy products. The dairy veterinarian and farmer have, for decades, thus been faced with the dual challenges of producing quality milk from healthy cows whilst not contaminating the food chain with medicine residues.

To make this possible, in many countries of the world, all AMs are prescription-only medicines, and their manufacture and supply are tightly regulated. Within the European Union (EU), for example, all medicines authorised for administration to a food-producing animal must have an established maximum residue limit and be listed in the table of allowed substances in the relevant EU legislation (Commission Regulation (EU) No 37/2010 http://ec.europa.eu/health//sites/health/files/files/eudralex/vol-5/reg_2010_37/reg_2010_37_en.pdf). No medicinal product may be administered to an animal unless an appropriate maximum residue limit is available in at least one food-producing species and a meat and milk withhold time has been established for the medicine in question.

In the veterinary sphere, each and every medicine is subject to tight regulatory control, and medicines are only licensed for use after detailed scrutiny of efficacy and safety data. For example, in the United Kingdom this is carried out by the

Veterinary Medicines Directorate's (VMD) Veterinary Products Committee (https://www.gov.uk/government/organisations/veterinary-products-committee). This results in each medicine licensed for administration within the EU having a summary of product characteristics (SPC), detailing, amongst other things, the therapeutic indications, methods of administration and food product withhold periods as well as other safety data and details of contraindications. This means that individual medicines are licensed for particular conditions and indications in specific animal species and for specified dose rates and therapeutic courses. For example, an intramammary AM product might be licensed as a single dose treatment for clinical mastitis in lactating cows, with a stipulation that milk be withheld from the supply chain for at least 72 hours and meat to be withheld for two days after the last administration. Deviation from the SPC is allowed in some countries under certain circumstances; for example, the United Kingdom has the 'Cascade' system, a risk-based decision tree that allows veterinarians to prescribe to animals under their care based upon their clinical judgement with the animal owner's consent as detailed below.

To quote the VMD's guidance on the cascade,

> 'If there is no suitable veterinary medicine authorised in the UK to treat a condition in a particular species, you can treat an animal under your care in accordance with the Cascade. The steps, in descending order of suitability, are:
>
> a veterinary medicine authorised in the UK for use in another animal species, or for a different condition in the same species
>
> If there is no such product, either: a medicine authorised in the UK for human use, or
>
> a veterinary medicine not authorised in the UK, but authorised in another member state for use in any animal species in accordance with the Special Import Scheme; in the case of a food-producing animal the medicine must be authorised in a food producing species
>
> a medicine prescribed by the vet responsible for treating the animal and prepared especially on this occasion (known as an extemporaneous preparation or special) by a vet, a pharmacist or a person holding an appropriate manufacturer's authorisation (so called specials manufacturers)
>
> in exceptional circumstances, medicines may be imported from outside Europe via the Special Import Scheme (VMD Guidance https://www.gov.uk/guidance/the-cascade-prescribing-unauthorised-medicines)

When a medicine is used in this way under the cascade rules (sometimes termed 'off-label'), there may be no predetermined milk and meat withhold period stipulated. In this case, a minimum of 7 days for milk and 28 days for meat must elapse between the last administration of the medicine and the harvesting of milk for human consumption or the slaughter of the animal for meat.

These controls aim to ensure only appropriate, efficacious medicines are administered to food animals and that public health is protected by preventing medicine residues in milk and meat. Farmers are educated in these areas by food chain actors and veterinarians to ensure medicine residues do not enter the food chain (Fig. 1). Alongside frequent and stringent testing of animal products to ensure residues do not reach the consumer, these regulations are very effective in protecting the animal against ineffective treatments and in ensuring there are no potentially harmful medicine residues in food. However, these regulations are not designed to safeguard against other potential risks such as AMs entering the environment or the development of antimicrobial resistance (AMR).

It should also be noted that in some parts of the world less stringent regulation is in place with not all AMs being prescription-only medicines; this poses a widespread risk given globalisation, rapid international travel and extensive trade of livestock and their products.

Many different pharmaceuticals are used in dairy cattle husbandry including vaccines, local anaesthetics, non-steroidal anti-inflammatory drugs (NSAIDs), hormones, anthelmintics, AMs and immune modulators, along with a diverse range of other products with specific indications. Whilst in most cases efficacy and safety are the primary considerations of the veterinarian when prescribing, in the case of anthelmintics and AMs, pathogen resistance is also an increasingly important consideration. Anthelmintic resistance is a significant animal health concern on which much has already been written,

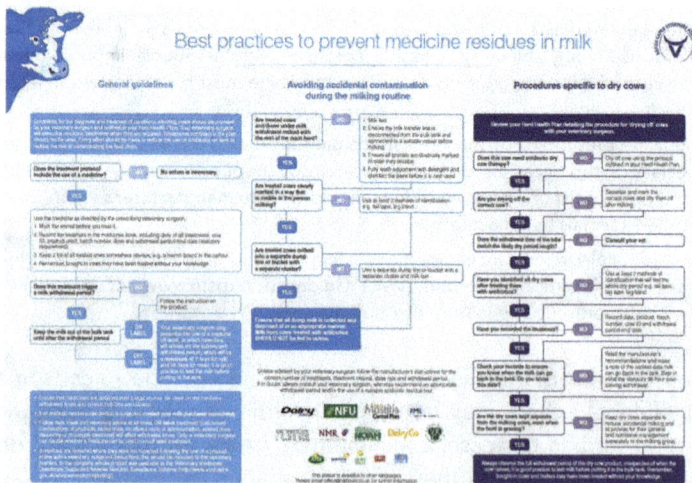

Figure 1 Poster illustrating the best practice to prevent medicine residues in milk, produced by the British Cattle Veterinary Association (BCVA) in conjunction with other industry stakeholders and distributed to all UK milk producers (also available in multiple languages from https://www.bcva.eu/resources/medicine-residues-milk-guidance-poster-0) (reproduced with the kind permission of the BCVA).

for example by the organisation Control of Worms Sustainably (COWS; http://www.cattleparasites.org.uk/), hence this chapter will not address that topic further.

AM/antibiotic (AB) resistance is a cross-species phenomenon that has the potential to have devastating consequences for both animal and human health. The dairy sector has a moral and ethical responsibility to address this threat, and there is an urgent need to adjust our husbandry and medicine prescribing practices to reduce this risk. Much of what was considered normal acceptable practice even a few years ago is now being increasingly scrutinised and challenged, and some uses of medicines will certainly be prohibited by many milk processors and major retailers in the future or even be banned by legislation.

2 Antimicrobial resistance

The rapid emergence and spread of AMR is a critical public health and One Health issue worldwide (Robinson et al., 2016). AMR has very profound implications not only for veterinary medicine but also for human health and global food security. The use of AMs in livestock is often implicated as influencing the amount of resistance present in the human population, and, while there remains much debate about whether such 'zoonotic transmission' happens to any significant degree, the agricultural and veterinary sectors must act responsibly in the way we use medicines now and in the future. There have been calls at the highest international political fora (including the G8, G20 and United Nations in 2016) for immediate and decisive action to be taken in human medicine, veterinary medicine and indeed anywhere AMs are used in agriculture, aquaculture or industry. Although there remain many questions about the real risk of AMR generated by use of AMs in cattle affecting the human population - either through direct contact with animals, through the food chain or the environment - there is more than enough evidence to support the need to take decisive action now to make sustainable changes to the way we use medicines.

AMR is a natural phenomenon, but AMs in some cases promote the *de novo* development of resistance and in every case select for bacteria able to survive in the presence of the AM. Whether they are pathogens or not, bacteria exposed to AMs - whether in an animal or the environment - will be driven by natural selection to evolve or acquire resistance mechanisms. The detail of these mechanisms are discussed elsewhere; what is important is to realise that all bacteria exposed to AMs/ABs are potentially capable of developing or acquiring resistance, and resistance to multiple AMs can be transferred between bacteria of the same and different species! All selection for resistance in all bacteria is therefore a risk to health, both animal and human, and even those ABs not considered critical for human health may co-select for resistance to critical last-resort medicines where multiple resistance genes are coded on the same plasmid.

Even if resistant bacterial pathogens are not causing disease on a farm, any use of AMs will increase the risk of *de novo* resistance developing or of resistance genes being selected within the microbiome of cattle or their environment. It follows, therefore, that the livestock industries need to dramatically reduce the amount of AMs given to animals. However, care needs to be taken to maintain animal health and welfare. Diseased animals require effective treatment, including, when appropriate, AMs, and simply stopping the use of medicines or curtailing how they are used by legislation will certainly have unintended welfare consequences.

In 2016, Lord Jim O'Neill and his team delivered 'The review on antimicrobial resistance' (O'Neill, 2015, 2016), and detailed recommendations for a significant increase in regulatory oversight of veterinary ABs, with the potential suggestion that veterinarians and farmers should benchmark levels of AM use and take action when consumption of ABs increases or exceeds accepted norms and targets. This report also recognised the need to compare use of AMs across the United Kingdom and Europe, and especially highlights the need to reduce, or even to ban, the use of ABs critical to human medicine in the livestock sector.

While the dairy sector uses less ABs than some other livestock sectors (both in real terms and when adjusted for animal biomass), dairy farms use relatively large amounts of the so-called critically important AMs (CIAs) – those considered critical to human health (ESVAC, 2016). This makes it imperative to focus on responsible use within this sector.

While there remains some debate about which ABs commonly used in bovine medicine should be considered as CIAs – for instance, the World Health Organisation and the European Medicines Agency have come to different conclusions with respect to macrolides (NOAH, 2017) – it is generally accepted that the use of third- and fourth-generation cephalosporins, fluoroquinolones and macrolides (especially the modern long-acting macrolides) are a particular concern (Table 1). In addition, colistin is considered a CIA but is not commonly used in cattle, at least in the United Kingdom. In future other AMs such as amoxicillin/clavulanic acid combinations may also be added to the list of CIAs for which there is particular concern, and veterinarians and herd owners should consider this when devising treatment protocols.

Limiting use of CIAs within the dairy sector raises specific challenges; for example, ceftiofur (a third-generation cephalosporin) in particular has a zero milk withhold in many countries, making its use in lactating cattle not only financially attractive to dairy producers but also provides an option to reduce

Table 1 Antimicrobials deemed critically important for human medicine which are currently licensed for use in cattle in many countries worldwide. Classes where there is particular concern over current use in food-producing animals are marked with an asterisk (*) (modified from Tisdall et al., 2017)

Class		Products used in cattle
Cephalosporins*	Third generation	Ceftiofur, cefoperazone
	Fourth generation	Cefquinome
Fluoroquinolones*		Danofloxacin, enrofloxacin, marbofloxacin
Macrolides*		Gamithromycin, tildipirosin, tilmicosin, tulathromycin, tylosin
Penicillins		Ampicillin, amoxycillin, penicillin G, penethamate hydroiodide
Aminoglycosides		Dihydrostreptamycin, framycetin, kanamycin, neomycin, streptomycin

the risk of food chain contamination. Similarly, the long-acting macrolides such as tulathromycin or gamithromycin are used extensively in the treatment (and sometimes in the control) of bovine respiratory disease (BRD) in young stock and growing heifers. It is essential, however, that the use of these CIAs is reduced as a matter of urgency across the dairy industry if we are to act responsibly. We need to focus both on overall AM use reduction targets – at the farm, veterinary practice and national levels – while at the same time driving the use of CIAs as low as possible if we are to reduce the risk of AMR on farms, and with it the risk to human health.

These authors are of the opinion that third- and fourth-generation cephalosporins and fluoroquinolones are not needed on dairy farms, or at least can be reserved for the exceedingly rare occasions when laboratory diagnostics show categorically that no other viable treatment is available. Macrolide use can also be substantially replaced by the use of other treatments along with good BRD-preventive strategies involving optimised husbandry and the strategic use of vaccination.

3 Inappropriate behaviours and practices

Although practices differ between countries and some, for example, only allow veterinarians to administer AMs to animals, others are less stringent in their controls. There are thus some commonplace behaviours on many dairy farms around the world that need to change, for example:

- Treatment decisions should always be made by a veterinarian or in direct communication with a veterinarian and not made by farm staff.

- AMs should be selected based on expected efficacy where possible supported by laboratory diagnostic and surveillance data. The risk of AMR developing or being selected for should always be considered; AMs should not be chosen solely based upon factors such as milk withhold periods.
- Conditions unlikely to be caused by bacteria should not be treated with ABs. For example, ABs are contraindicated in most diarrhoeic calves, since most diarrhoea has a viral cause and ABs alter the normal, protective gut microflora. Likewise mild BRD may be better treated with a NSAID than with an AB in many cases.
- Feeding waste milk contaminated with low, sub-therapeutic doses of ABs to young calves, while perhaps reducing calf-rearing costs, is clearly a significant risk for AMR development and should no longer be practiced (Brunton et al., 2012, 2014; Duse et al., 2015; Ricci et al., 2017). Such milk should be discarded and not fed to livestock of any type.
- Wherever possible non-antibiotic udder protectant teat sealants should be used rather than ABs at the end of lactation (Scherpenzeel et al., 2014; Biggs et al., 2016). Not only does this substantially reduce AB use on farms practicing such selective dry cow therapy, it also reduces the amount of colostrum containing low dose antibiotics fed to calves.
- Feeding milk powder with ABs included to calves as a prophylactic measure should be ceased.
- Using footbaths containing ABs to treat, or control, infectious foot conditions, particularly digital dermatitis should be curtailed. Better on-farm hygiene and appropriate use of non-AB footbaths need to be employed rather than over-reliance on ABs. For example, Solano et al. (2017) have shown that appropriate use of non-AB footbath solutions such as copper and zinc salts, or formalin, can be used to control digital dermatitis. While far better than using ABs, these methods are not risk-free, however, with respect to AMR, as the heavy metals have also been implicated in selection for resistant organisms in certain situations.

Many global organisations have called for change and more responsible use of medicine in agriculture, including the World Health Organization, the Food and Agriculture Organization of the United Nations (FAO) and the World Organisation for Animal Health (OIE).

With particular relevance to dairy farms in the United Kingdom, the British Veterinary Association (BVA) (Fig. 2), British Cattle Veterinary Association (BCVA), Responsible Use of Medicines in Agriculture Alliance (RUMA http://www.ruma.org.uk/), Dairy UK and the British Society for Antimicrobial Chemotherapy (BSAC) (Fig. 3) have all called for change. The Dairy UK and

Figure 2 British Veterinary Association's (BVA) poster illustrating responsible use of antimicrobials in veterinary practice (reproduced with the kind permission of the BVA).

BCVA farmer training initiative 'MilkSure' has been specifically designed to increase awareness of AMR amongst milk producers in the United Kingdom (Milksure http://milksure.co.uk/).

4 Making progress towards change

Although some veterinarians have shown clear progress in adjusting prescribing practices and even stopping the use of CIAs on dairy farms altogether (Tisdall et al., 2016), there are certain challenges to overcome. To begin with, veterinarians are professionals who, in many countries, directly benefit financially from the sale of AMs, a scenario which presents them with a potential conflict of interest. Like medical professionals, veterinarians are often risk-averse, and might feel pressured by farmers to prescribe ABs, even when there is little or no evidence that farmers themselves are demanding prescriptions (Coyne et al., 2014). Some farmers have even reported feeling that their veterinarian might disapprove of them decreasing their AB usage (Jones et al., 2015). Farmers also frequently report that their veterinarians are not engaging them on the issue of AM stewardship, and while veterinarians recognise their influence and the need for them to be proactive advisors, making positive progress can still be a challenge (Buller et al., 2015).

We recommend prevention of overuse of antibiotics in all animals to be achieved by:

- A global ban on the use of all antibiotic growth promoters.
- Minimizing prophylactic and metaphylactic use of antibiotics in all animals.
- Enhanced surveillance of animals for both animal and human infectious and commensal bacteria so as to provide an evidence base on resistance genes and their transmission within animal populations, and between animals and humans.
- Employing effective infection control measures to prevent dissemination of resistance genes between animals and humans, and within and between animal populations (e.g. farms, kennels, etc).
- Increased research investment into animal husbandry and disease prevention in livestock and aquaculture systems.
- Fluoroquinolones, 3rd and 4th generation cephalosporins, and colistin only to be used in animals in exceptional circumstances and only after showing a definite need.
- Not using any new classes of antibiotic in animals; reserving them for human use only, unless they are found to only be safe in animals, not used in people and do not cause resistance to medicines used in people.
- Increased monitoring of animal husbandry practices through enhanced quality assurance schemes, and the promotion of best practice to optimize animal welfare and minimize the need for therapeutic medication in livestock.
- Food buyers, processors and retailers to promote evidence-based best practice in animal welfare and the use of antibiotics in all parts of their supply chains. They should provide information on packaging to allow consumers to identify produce from farm assured supply chains.
- Enhance data collection on antibiotic prescribing by all veterinary practitioners and usage of antibiotics on every farm. Consider benchmarking of antibiotic use at the farm level and publishing benchmarking data on open-access websites.
- Funding for research into diagnostic techniques for bacterial infections in animals should be increased.
- Vaccination to become the primary means of preventing animal infectious diseases especially in agriculture and aquaculture. Increased funding is vital to facilitate research into the development and delivery of new vaccines.

Figure 3 Policy statement by the British Society for Antimicrobial Chemotherapy (BSAC) on the use of antibiotic medication in animals published May 2016.

As an example of what can be achieved, the University of Bristol Farm Animal Practice have shown that very substantial and lasting changes can be made to medicine prescribing by integrating a holistic medicine use review process into herd health management. This strategy focused on all medicine use, not just AMs. For example, auditing NSAIDs and local anaesthetic use can provide valuable insight into animal welfare and pain management, while considering vaccination usage can give very valuable insight into whether attitudes are geared towards preventive or reactive approaches to herd health (Tisdall et al., 2017). Integrating a detailed review of actual medicine use on-farm into health planning in this way means that measures can be taken to improve animal health, reducing the numbers of animals treated, as well as ensuring that when treatments are required they are applied appropriately. This approach can greatly enhance the farmer-veterinarian working relationship, making preventive medicine a reality and enhancing animal health while reducing both medicine costs and the risk of AMR.

In 2010, veterinarians at the University of Bristol's Farm Animal Practice began an initiative to reduce the use of CIAs on all farms under the care of the practice. During 2010-11 farmers were engaged and educated about this process through farmer meetings and visits to discuss herd health planning; during 2011-12 changes to prescribing policy were implemented on pilot farms. This initiative was then extended to all farms through 2012-15. Farmers worked closely with different veterinarians during this time, but the same changes to prescribing practice were encouraged on all farms throughout the process. A strict prescribing protocol for 'Protected Antimicrobials' – third- and fourth-generation cephalosporins, fluoroquinolones and longer-acting macrolides – was also adopted by all veterinarians in the practice (Fig. 4; Tisdall et al., 2016; Turner et al., In Press).

This approach required an integration of improved herd health management, clinical governance and evidence-based veterinary medicine (EBVM) along with good communication and a team approach to bring about more responsible medicine use on farms and a continuous and ongoing cycle of progressive change and continuous monitoring and evaluation.

A continuous process of review and adaptation based upon a collaborative dialogue between the veterinarian and the farmer was required to bring about lasting behavioural change. A mistake that is often made by veterinarians is to assume farmers are resistant to change, while farmers may feel veterinarians are not challenging them and providing them with the assistance they need to change their medicine usage. Communication methodologies such as motivational interviewing or participatory approaches such as peer group learning, alongside benchmarking, can be particularly beneficial in facilitating change (Van Dijk et al., 2017; Reyher et al., 2017). Whether on one farm or a group of farms, the process needs to be one of ongoing dialog and challenging the status quo. This is a continuous cyclical process which hopefully leads to long-term change (Table 2).

The following principles are important to get the full benefit of any initiative to improve medicine usage on farm (Tisdall et al., 2017):

Langford Farm Animal Practice
Protocol for the use of Protected Antimicrobials

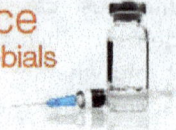

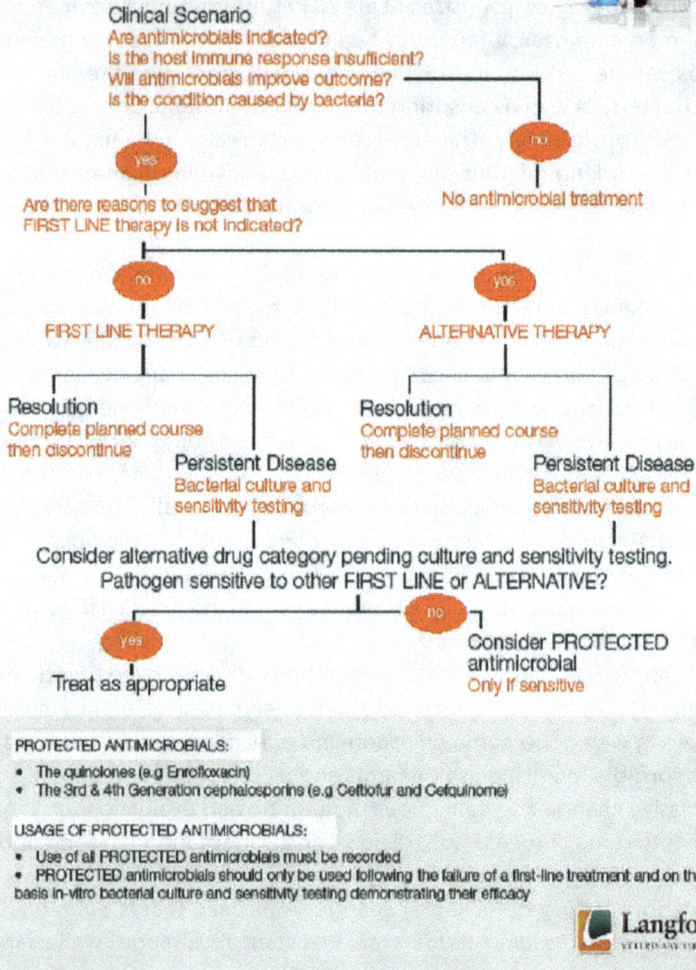

Clinical Scenario
Are antimicrobials indicated?
Is the host immune response insufficient?
Will antimicrobials improve outcome?
Is the condition caused by bacteria?

yes

no

Are there reasons to suggest that
FIRST LINE therapy is not indicated?

No antimicrobial treatment

no

yes

FIRST LINE THERAPY

ALTERNATIVE THERAPY

Resolution
Complete planned course
then discontinue

Persistent Disease
Bacterial culture and
sensitivity testing

Resolution
Complete planned course
then discontinue

Persistent Disease
Bacterial culture and
sensitivity testing

Consider alternative drug category pending culture and sensitivity testing.
Pathogen sensitive to other FIRST LINE or ALTERNATIVE?

yes

no

Treat as appropriate

Consider PROTECTED
antimicrobial
Only if sensitive

PROTECTED ANTIMICROBIALS:
- The quinolones (e.g Enrofloxacin)
- The 3rd & 4th Generation cephalosporins (e.g Ceftiofur and Cefquinome)

USAGE OF PROTECTED ANTIMICROBIALS:
- Use of all PROTECTED antimicrobials must be recorded
- PROTECTED antimicrobials should only be used following the failure of a first-line treatment and on the basis in-vitro bacterial culture and sensitivity testing demonstrating their efficacy

Langford
VETERINARY SERVICES

Figure 4 University of Bristol Farm Animal Practice protocol for use of critically important antimicrobials. The term 'protected' antimicrobials here refers to third- and fourth-generation cephalosporins, fluoroquinolones and the longer-acting macrolides (Tisdall et al., 2017).

4.1 SMART (Specific, Measurable, Achievable, Relevant and Time-bound) objectives

The objectives should be farm-specific, although they might be informed by wide-reaching initiatives such as national targets or targets set by individual

Table 2 Tools to help farmers and veterinarians work together towards more responsible antimicrobial use (modified from Reyher et al., 2017)

What we can do	How we can do it
Improving veterinary-farmer communication	Foster collaboration, ask and listen (more than talking!), co-create knowledge, use the farmer's bespoke knowledge of their own farm and farming practices
Participatory approaches	Actively involve farmers in devising responsible medicine use strategies for their farms, share best practice between farms
Accurate records of medicines use	Use easy methods of data recording and analysis
Benchmarking	Benchmark farms and veterinarians to compare AM use, to motivate discussion and change
Active herd health management	Establish herd health management processes that are fit-for-purpose, specifically focusing on medicines (or specific medicine) use, and useful on-farm protocols
Education and training	Better veterinarian and farmer training on responsible AM use. Train everyone involved on the farm, especially those making treatment decisions
Improve the system	Work together to understand the culture and practices on each farm and what the barriers to change might be

milk buyers or retailers. These objectives need to be SMART, for example, a 20% reduction in third-generation cephalosporin use in three months might satisfy all of these criteria, but so might an interim goal such as to undertake a two-hour discussion with all staff on the farm to exchange ideas and explore options for change within the next two weeks. Defining desired SMART outcomes allows progress to be demonstrated, recognised, encouraged and rewarded. One should start small and build momentum over time, since objectives will inevitably change as progress is made and more ambitious targets are set. If progress is slow, or fails to be achieved, this process also allows reflection and goal re-evaluation.

4.2 Cultural change and strong leadership

Both at the farm and veterinary practice level this will be important if lasting change is to be achieved; the *status quo* needs to be challenged and each individual's natural reluctance to change, or fear of change, overcome. A wider adoption of EBVM principles, for example, using the RCVS Knowledge EBVM Toolkit (2017) (http://knowledge.rcvs.org.uk/evidence-based-veterinary-medicine/ebvm-toolkit/) may go hand-in-hand with this approach, as the cultural norm becomes science-driven and less about personalities and opinion which are traditionally based on anecdote and seniority and may be unsubstantiated by evidence.

4.3 Widespread involvement

Engendering behaviour change of this type needs *all* those involved in the care and treatment of animals on the farm, or within the veterinary practice (including support staff) to take ownership of responsible medicine use and to understand and commit to change.

4.4 Collaborative participation

The team at the University of Bristol found that one of the key factors for success in achieving behavioural change towards more responsible medicine use amongst both farmers and veterinarians was the adoption of a collaborative approach (Tisdall et al., 2016). Farmers and farm staff were drawn in as partners in the process, and veterinary practice policy and behavioural change moved forward together. This also involved a concerted and targeted education programme, but one-on-one, on-farm conversations as part of herd health planning helped to make the changes sustainable.

Similar processes driven by government enforced targets have been employed successfully in the Netherlands to bring about behavioural change and a substantial reduction in AM usage in livestock farming (Speksnijder et al., 2014).

4.5 Medicine auditing and clinical governance

In order for AM use to be considered responsible it must be continually measured, monitored and reviewed alongside all other aspects of herd health. Systematic farm- and practice-level medicine auditing is a useful way to achieve this. These audits identify key opportunities for further improvement and allow benchmarking of progress. It is likely that in many countries farm-, veterinary practice and even veterinarian-level audits and benchmarking will become the future norm if we are to avoid further regulation and legislation restricting prescribing.

Medicine recording and auditing software tools and apps are likely to become increasingly available in the future, with many in development at the time of writing (May 2017); for example, the veterinary antibiotic use calculator developed by LEI Wageningen UR in the Netherlands is quite useful (https://www3.lei.wur.nl/antibiotica/Default.aspx).

4.6 Sustainability

It is essential to develop a culture in the veterinary practice and on-farm that will last. To be sustainable, change must extend beyond the influence of a small number of individuals who may have been pivotal in the introduction of new approaches and protocols. Establishing agreed protocols that everyone involved signs up to will help, but habituating both veterinary and farm staff to an ongoing process of training, discussion and retraining is key. Policies on medicine use and specific training on AM use should form part of the induction process of all involved in medicine prescribing or animal care on a farm, as well as support staff within veterinary practices. Responsible medicine use must become the cultural norm and underpin every aspect of clinical practice and herd health management on every farm.

Using a holistic approach such as this has been shown to be very effective in bringing about progressive and lasting change in medicine use within the University of Bristol Farm Animal Practice (Tisdall et al., 2016, 2017), as well as large commercial farm animal veterinary practices in the United Kingdom who have followed all, or part, of this model. Similarly, work in other countries has shown downward trends in exposure of farm animals to antibiotics, such as the MARAN (Monitoring of AMR and antibiotic usage in animals in the Netherlands) initiative in the Netherlands (http://www.wur.nl/en/Research-Results/Projects-and-programmes/MARAN-Antibiotic-usage/Introduction.htm). Changes in AM use in animals at a national level are now regularly monitored, for example the European Commission reports on AM use in Denmark and the Netherlands (European Commission 2016, 2017) and the EU-wide monitoring reported annually by the European Medicines Agency (e.g. ESVAC, 2016).

5 Delivering results

Continuing the case study described above, the University of Bristol Farm Animal Practice undertook a practice-wide policy change in 2010 which resulted in the following specific AM prescribing changes being implemented throughout the practice to reduce and ultimately eliminate the use of CIAs on dairy farms (Tisdall et al., 2016; Turner et al., In Press).

- **Intramammary use** – Fourth-generation cephalosporins were phased out in favour of penicillin and aminoglycoside combinations.
- **Systemic use** – Fluoroquinolone use had been halted in 2009, and third- and fourth-generation cephalosporins were gradually replaced with a first-generation cephalosporin or aminopenicillins.
- **Calf pneumonia** – Longer-acting macrolides were replaced with oxytetracycline or florfenicol and a greater emphasis was placed on non-AM therapies (e.g. NSAIDs).

The medicine audit and health data, along with farmer feedback, show categorically that these changes can be made without any perceived or actual detrimental effects on animal health or productivity (Fig. 5–7).

Throughout the whole of the Farm Animal Practice, CIA use was radically reduced with an 82% reduction overall, a 91% reduction in systemic use and a 100% reduction in intramammary use being achieved over a six-year period (2010–2015), with the majority of that reduction occurring in the first two years after the process was initiated. The use of CIAs approached zero in 2016, with no CIAs having been used in intramammary products for three years (Fig. 5).

Over the same six-year period, cow health, welfare and production have been maintained or improved. For example, data from five dairy farms are shown in Fig. 6 and 7. Analysis of health data from dairy farms across the practice demonstrates that udder health in particular improved over the six years during which CIA use was reduced, with the mean case rate of clinical mastitis decreasing from 34 to 21 cases/100 cows/year. Over the same period the subclinical infection rate remained relatively consistent, with the mean proportion of cows with SCC above 200 000 cells/ml being between 17 and 20%, while the mean 305-day yield increased slightly from 8600 to 9000 L (Fig. 6).

Lameness, as defined by AHDB Mobility Scoring (2017), throughout these five farms also improved despite the fact that treatment protocols involving third- and fourth-generation cephalosporins were replaced by aminopenicillins or a first-generation cephalosporin product (in cases where a zero milk withhold was required). Although there are of course many risk factors for lameness other than infectious conditions, it is noteworthy that the mean levels of lameness recorded on farms for which lameness data was available in 2010/11 were 29% compared with 15.5% in 2015 (Fig. 7).

Research looking just at dairy farms within the University of Bristol's Farm Animal Practice showed that during the six years of the study, prescriptions of

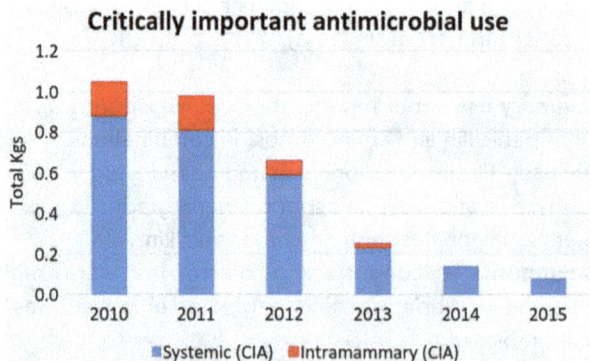

Figure 5 Fall in critically important antimicrobial (CIA) use in the whole of the University of Bristol Farm Animal Practice – representing 82% reduction in overall CIA use, 91% reduction in systemic use and 100% reduction in intramammary use.

Udder health on five dairy farms

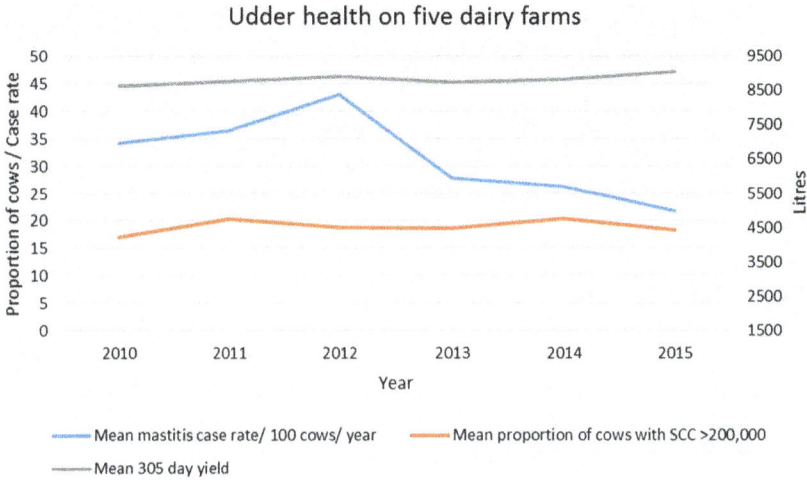

Figure 6 Udder health data for five dairy farms during the period that use of CIAs were totally phased out on these farms (2010-2015). Data show a fall in clinical mastitis cases, stability in the proportion of cows with SCC above 200 000 cells/ml and an increase in milk yield over this period.

CIAs decreased from 0.58 average daily doses (ADD)/animal/yr (35% of ADD/animal/yr in 2010) to 0 ADD/animal/yr in 2015 (Turner et al., In Press; ADD as defined by Dupont et al., 2016). Intramammary preparations containing CIAs had been discontinued by 2014.

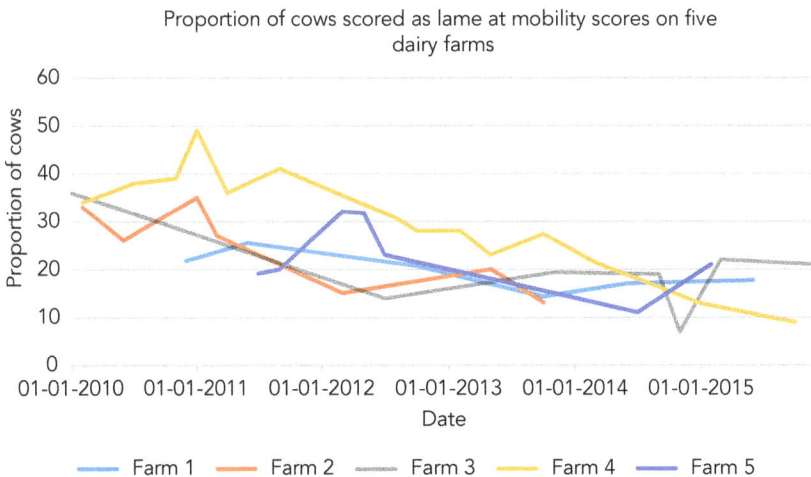

Figure 7 Lameness data for five dairy farms during the period that use of CIAs were totally phased out on these farms (2010-2015). Data show a general decrease in number of cows classified as lame using the AHDB Dairy Mobility Scoring system (lame cows scored 2 and 3).

Production parameters including 305-day yield, milk protein and butterfat percentages remained stable throughout the study (Turner et al., In Press). Most health parameters improved: weighted mean (W-M) calving index decreased from 415 to 388 days, W-M calving to conception interval decreased from 152 to 119 days and W-M 100 day in calf rates increased from 36.2 to 47.2%. Clinical mastitis case rates decreased on most farms, and clinical mastitis cure rates increased on the majority of farms (30 to 42%). Mobility scores indicated lameness rates on all of the study farms decreased over the study period, and culling rates remained stable (Turner et al., In Press).

In summary, research at the University of Bristol has demonstrated that dairy cattle health and welfare – as measured by production parameters, fertility, udder health, mobility data and culling rates – can be maintained and even improved substantially alongside a complete cessation in the use of CIAs as well as an overall reduction of all AM use on dairy farms. Furthermore, there was no perception amongst veterinarians, or farmers, that treatment success rates reduced during this transition to more responsible medicine use, an important consideration when promoting behavioural change.

Similarly impressive progress in responsible medicine prescribing has also been achieved by large commercial veterinary practices in South West England working in collaboration with the University of Bristol's AMR Force research group (@AMRForce). Change is possible and not as difficult to achieve as some might think!

There is a compelling need for changes in medicine use, to reduce overall AM usage and phase out CIAs altogether in dairy production. The University of Bristol group has shown this can be achieved alongside improved animal health and welfare. The challenge remains for the dairy industry at large to follow this lead and to move to sustainable treatment and disease prevention protocols. With improved animal health, veterinarians and farmers will not require AMs considered to be critically important for human health, and in so doing will reduce as far as possible the risk of AMR developing on dairy farms. All of this can be done alongside safeguarding animal health, animal welfare, the food chain and the environment.

6 Future trends and conclusion

Reduced use of AMs and better use of other medicines on dairy farms is readily achievable if all involved in delivering health care and managing the husbandry of the herd work together towards this goal. What was once considered legitimate use of medicines in the dairy sector might now be considered abuse of the veterinarians' right to prescribe. We are already seeing retailers and milk buyers pushing hard for real change in medicine usage at the farm level, and in some countries national reduction targets have been in place for some years. Our use of medicines in dairy production needs to change and become far more sustainable if we are to avoid further regulation or legislative bans on certain medicines or ways of using medicines. Safeguarding animal health and the food chain from residues is no longer sufficient; veterinarians need to consider the bigger picture of the environment and AMR.

7 Where to look for further information

Further more detailed information on applied AMR research at the University of Bristol can be found in the following:

Reyher, K.K., Barrett, D.C. and Tisdall, D.A. (2017) Achieving responsible medicine use: communicating with farmers. *In Practice* 39, 63–71 doi: 10.1136/inp.j341.

Tisdall, D.A., Reyher, K.K. and Barrett, D.C. (2017) Achieving responsible medicines use at practice and farm level. *In Practice* 39, 119–127 doi: 10.1136/inp.j658.

8 Acknowledgements

The authors wish to acknowledge the partnership of staff and students involved in the University of Bristol's AMR Force (@AMRForce). This work would not be possible without the excellent cooperation and collaboration between those in this multidisciplinary research group and their collaborators within and outside the University of Bristol. Particular thanks are due to Dr Jude Capper for her assistance in drafting some sections and to the publishers of the journal *In Practice*; BMJ, BMA House, London, WC1H 9JR, since parts of this chapter are based around materials published in two 'opinionated reviews' from the journal.

9 References

AHDB Dairy Mobility Scoring (2017). https://dairy.ahdb.org.uk/technical-information/animal-health-welfare/lameness/husbandry-prevention/mobility-scoring/#.WNLX9FXyjRZ Accessed 22 March 2017.

Biggs, A., Barrett, D. C., Bradley, A., Green, M., Reyher, K. and Zadoks, R. (2016). Antibiotic dry cow therapy: where next? *Vet. Rec.* 178, 93–4. http://dx.doi.org/10.1136/vr.i338.

Brunton, L. A., Duncan, D., Coldham, N. G., Snow, L. C. and Jones, J. R. (2012). A survey of antimicrobial usage on dairy farms and waste milk feeding practices in England and Wales. *Vet. Rec.* 171, 286. http://dx.doi.org/10.1136/vr.100924.

Brunton, L. A., Reeves, H. E., Snow, L. C. and Jones, J. R. (2014). A longitudinal field trial assessing the impact of feeding waste milk containing antibiotic residues on the prevalence of ESBL-producing *Escherichia coli* in calves. *Prev. Vet. Med.* 117, 403–12. http://dx.doi.org/10.1016/j.prevetmed.2014.08.005.

Buller, H., Hinchliffe, S., Hockenhull, J., Barrett, D., Reyher, K., Butterworth, A. and Heath, C. (2015). Systematic review and social research to further understanding of current practice in the context of using antimicrobials in livestock farming and to inform appropriate interventions to reduce antimicrobial resistance within the livestock sector. Research Report, Defra. file:///C:/Users/lvdcb/Chrome%20Local%20 Downloads/12817_ReportO00558Final%20(4).pdf Accessed 22 March 2017.

Capper, J. L. and Yancey, J. W. (2015). Communicating animal science to the general public. *Anim. Front.* 5, 28–35. doi:10.2527/af.2015-0028.

CAST (2013). *Animal Feed vs. Human Food: Challenges and Opportunities in Sustaining Animal Agriculture Toward 2050*. Council for Agricultural Science and Technology: Des Moines, IA, USA. https://www.cast-science.org/publications/?animal_feed_vs_ human_food_challenges_and_opportunities_in_sustaining_animal_agriculture_tow ard_2050&show=product&productID=278268 Accessed 27 March 2017.

Commission Regulation (EU) No 37/2010 (2010). http://ec.europa.eu/health//sites/ health/files/files/eudralex/vol-5/reg_2010_37/reg_2010_37_en.pdf.

COWS (2017). Control of worms sustainably. http://www.cattleparasites.org.uk/Accessed 22 March 2017.

Coyne, L. A., Pinchbeck, G. L., Williams, N. J., Smith, R. F., Dawson, S., Pearson, R. B. and Latham, S. M. (2014). Understanding antimicrobial use and prescribing behaviours by pig veterinary surgeons and farmers: a qualitative study. *Vet. Rec.* 175, 593. http://dx.doi.org/10.1136/vr.102686.

Dupont, N., Fertner, M., Kristensen, C. S., Toft, N. and Stege, H. (2016). Reporting the national antimicrobial consumption in Danish pigs: influence of assigned daily dosage values and population measurement. *Acta Vet. Scand.* 58, 27. doi: 10.1186/ s13028-016-0208-5.

Duse, A., Waller, K. P., Emanuelson, U., Unnerstad, H. E., Persson, Y. and Bengtsson, B. (2015). Risk factors for antimicrobial resistance in fecal *Escherichia coli* from preweaned dairy calves. *J. Dairy Sci.* 98, 1–17. http://dx.doi.org/10.3168/jds.2014-8432.

ESVAC (2016). European Medicines Agency, European Surveillance of Veterinary Antimicrobial Consumption. 'Sales of veterinary antimicrobial agents in 29 European countries in 2014'. http://www.ema.europa.eu/docs/en_GB/document_library/ Report/2016/10/WC500214217.pdf Accessed 29 March 2017.

European Commission (2016). Final report of a fact-finding mission carried out in Denmark from 01 February 2016 to 05 February 2016 in order to gather information on prudent use of antimicrobials in animals. http://ec.europa.eu/food/audits- analysis/act_getPDF.cfm?PDF_ID=12500 Accessed 27 March 2017.

European Commission (2017). Final report of a fact-finding mission carried out in the Netherlands from 13 September 2016 to 20 September 2016 in order to gather information on the prudent use of antimicrobials in animals. http://ec.europa.eu/ food/audits-analysis/act_getPDF.cfm?PDF_ID=12902 Accessed 27 March 2017.

FAO (2015). *The State of Food Insecurity in the World*. FAO: Rome, Italy. http://www.fao. org/hunger/key-messages/en/Accessed 27 March 2017.

FAO (2017). *The Future of Food and Agriculture*. FAO: Rome, Italy. http://www.fao.org/ publications/fofa/en/Accessed 27 March 2017.

Jones, P. J., Marier, E. A., Tranter, R. B., Wu, G., Watson, E. and Teale, C. J. (2015). Factors affecting dairy farmers' attitudes towards antimicrobial medicine usage in cattle in England and Wales. *Prev. Vet. Med.* 121, 30–40. http://dx.doi.org/10.1016/j. prevetmed.2015.05.010.

Kearney, J. (2010). Food consumption trends and drivers. *Philos. Trans. R. Soc. B.* 365, 2793–807. doi: 10.1098/rstb.2010.0149.

LEI Wageningen UR (2017) in the Netherlands. https://www3.lei.wur.nl/antibiotica/Default.aspx.

MARAN (2017). http://www.wur.nl/en/Research-Results/Projects-and-programmes/MARAN-Antibiotic-usage/Introduction.htm Accessed 27 March 2017.

Milksure (2016). http://milksure.co.uk/Accessed 22 March 2017.

Neumann, C. G., Murphy, S. P., Gewa, C., Grillenberger, M. and Bwibo, N. O. (2007). Meat supplementation improves growth, cognitive, and behavioral outcomes in Kenyan children. *J. Nutr.* 137, 1119-23.

NOAH (2017). Critically Important Antibiotics in Veterinary Medicine: European Medicines Agency Recommendations. http://www.noah.co.uk/wp-content/uploads/2016/12/NOAH-briefing-on-CIAs-07122016.pdf Accessed 22 March 2017.

O'Neill, J. (2015). Antimicrobials in agriculture and the environment: Reducing unnecessary use and waste. https://amr-review.org/sites/default/files/Antimicrobials%20in%20agriculture%20and%20the%20environment%20-%20Reducing%20unnecessary%20use%20and%20waste.pdf Accessed 22 March 2017.

O'Neill, J. (2016). Tackling drug-resistant infections globally: Final report and recommendations. https://amr-review.org/sites/default/files/160525_Final%20paper_with%20cover.pdf Accessed 22 March 2017.

RCVS Knowledge – EBVM Toolkit (2017). http://knowledge.rcvs.org.uk/evidence-based-veterinary-medicine/ebvm-toolkit/Accessed 22 March 2017.

Ricci, A., Allende, A., Bolton, D., Chemaly, M., Davies, R., Escamez, P., Girones, R., Koutsoumanis, K., Lindqvist, R., Nørrung, B., Robertson, L., Ru, G., Sanaa, M., Simmons, M., Skandamis, P., Snary, E., Speybroeck, N., Kuile, B., Threlfall, J., Wahlstrom, H., Bengtsson, B., Bouchard, D., Randall, L., Tenhagen, B., Verdon, E., Wallace, J., Brozzi, R., Guerra, B., Liebana, E., Stella, P. and Herman, L. (2017). Risk for the development of Antimicrobial Resistance (AMR) due to feeding of calves with milk containing residues of antibiotics. *Eur. Food Saf. Authority J.* 15, 4665. doi: 10.2903/j.efsa.2017.4665. Accessed 27 March 2017.

Robinson, T. P., Bu, D. P., Carrique-Mas, J., Fèvre, E. M., Gilbert, M., Grace, D., Hay, S. I., Jiwakanon, J., Kakkar, M., Kariuki, S., Laxminarayan, R., Lubroth, J., Magnusson, U., Thi Ngoc, P., Van Boeckel, T. P. and Woolhouse, M. E. J. (2016). Antibiotic resistance is the quintessential One Health issue. *Trans. R. Soc. Trop. Med. Hyg.* 110, 377-80. doi:10.1093/trstmh/trw048.

RUMA - Responsible Use of Medicines in Agriculture Alliance (2017). http://www.ruma.org.uk/Accessed 22 March 2017.

Scherpenzeel, C. G. M., den Uijl, I. E. M., van Schaik, G., Olde Riekerink, R. G. M., Keurentjes, J. M. and Lam, T. J. G. M. (2014). Evaluation of the use of dry cow antibiotics in low somatic cell count cows. *J. Dairy Sci.* 97, 3606-14. http://dx.doi.org/10.3168/jds.2013-7655.

Solano, L., Barkema, H. W., Pickel, C. and Orsel, K. (2017). Effectiveness of a standardized footbath protocol for prevention of digital dermatitis. *J. Dairy Sci.* 100, 1295-307. http://dx.doi.org/10.3168/jds.2016-11365.

Speksnijder, D. C., Mevius, D. J., Bruschke, C. J. M. and Wagenaar, J. A. (2014) Reduction of veterinary antimicrobial use in the Netherlands. The Dutch success model. *Zoonosis and Public Health* 62, 79-87. doi: 10.1111/zph.12167.

Tisdall, D. A., Barrett, D. C. and Reyher, K. K. (2016). Developing multifaceted, collaborative, practice-wide approaches to responsible medicine use on farms. *The 29th World Buiatrics Congress*, 3rd-8th July 2016, Dublin, Ireland, pp. 129-30.

Van Dijk, L., Hayton, A., Main, D. C. J., Booth, A., King, A., Barrett, D. C., Buller, H. J. and Reyher, K. K. (2017). Participatory policy making by dairy producers to reduce anti-microbial use on farms. *Zoonosis and Public Health*. doi: 10.1111/zph.12329.

Veterinary Products Committee (2017). https://www.gov.uk/government/organisations/veterinary-products-committee Accessed 22 march 2017.

VMD Guidance (2017) https://www.gov.uk/guidance/the-cascade-prescribing-unauthorised-medicines Accessed 22 march 2017.

Chapter 2

Preventing bacterial diseases in dairy cattle

Sharif S. Aly[1] and Sarah M. Depenbrock[1], University of California-Davis, USA

1 Introduction

2 Pathogen host environment: an overview

3 Disease detection

4 Risk assessment tools

5 Future trends in research

6 Where to look for further information

7 References

1 Introduction

A paradigm shift in the management of bacterial diseases of food animals is needed; the focus should shift from pharmacologic interventions to disease risk identification and mitigation. Food animal management practices that focus on matching a pathogen with a pharmaceutical treatment (aka 'bug and drug') are oversimplified, threaten the sustainability of food animal production and erode consumer trust in animal source protein. Maintaining animal health, judicious use of antibiotics and sustainability of animal production systems are common expectations for food animal production. Relying on pharmaceuticals alone is insufficient to prevent serious disease syndromes associated with bacterial infections in cattle; indiscriminate use of antimicrobial drugs without careful adjustments to management practices results in treatment failures, animal suffering and greater risk for antimicrobial resistance (AMR), further comprising animal and public health (Oliver et al., 2011; Vaarten, 2012).

Dairy cattle production requires intense animal management systems that involve continuous daily interventions to harvest milk, increase fecundity and hand-raise young stock. Such systems allow for constant fine-tuning and the opportunity to increase both milk and dairy beef commodity production. However, these intense animal management operations also create multifactorial

1 Both authors contributed equally.

http://dx.doi.org/10.19103/AS.2020.0086.16

animal health risks. The interaction between disease risk factors can be nicely described by a conceptual model called the disease triangle. This model, as displayed in Fig. 1, incorporates the host, pathogen and environmental factors as a tool to understand different aspects that influence the disease. The disease triangle was first described by plant scientists, was later applied to public health and is now occasionally used in veterinary science (Scholthof, 2007; Woolums and Step, 2020). A sustainable approach to bacterial disease mitigation in production settings must include a comprehensive assessment of the host, pathogen and environmental factors of the disease. The authors propose that the continuum of a disease between the host, environment and pathogen is so inextricably linked that they cannot be separated when discussing cause versus prevention of disease. This chapter will focus on the state of the art of disease prevention in dairy cattle, specifically focusing on preventing bacterial causes of bovine respiratory disease (BRD). Each of the chapter sections identifies mechanisms of the disease triangle that can be disrupted to prevent diseases; these concepts culminate in a specific risk assessment tool in the third section of the chapter that can be used to determine a farm-specific action plan for disease prevention using calfhood respiratory disease as an example. Through the risk assessment approach, investigations into interactions between the host, environment and pathogen can provide robust information on which to build a prevention platform for multiple syndromes of bacterial diseases in

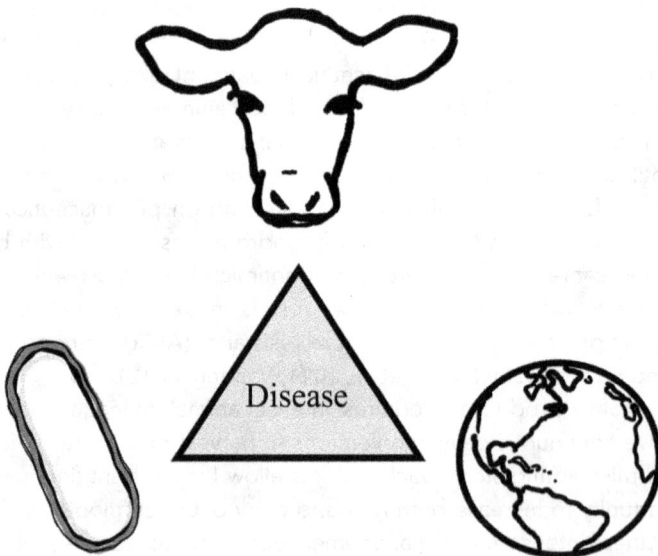

Figure 1 Schematic diagram of host (calf), environment (globe) and pathogen (bacterial rod) interactions. Original artwork by Dr. Sarah Depenbrock.

cattle. While we chose BRD as the working example, the same principles can be applied to other diseases such as neonatal calf diarrhea, which shares several risk factors with respiratory disease in calves.

1.1 BRD as the focus of disease prevention in dairy cattle

Arguably one of the best examples of the disease triangle in cattle is the BRD complex. The BRD complex is a syndrome of respiratory disease that results from the complex interactions between the host, environment and pathogen and culminates in a bacterial infection of the lower airways. The BRD complex was selected for discussion due to the relatively active areas of research in all three aspects of the disease triangle, the significance of BRD in the cattle industry and the overlap in preventive strategies between BRD prevention and prevention of other syndromes of bacterial diseases in dairy cattle. Like many diseases of dairy cattle, BDS does not often occur in isolation; underlying problems with one or more aspects of the disease triangle may predispose cattle to BRD. Many of the strategies discussed in the context of BRD prevention rely on maximizing host immunity, reduction of environmental stressors and contamination and, where appropriate, specific pathogen treatment and immunoprophylaxis. Although the focus of this chapter is BRD, most of these prevention strategies are common to the prevention of other diseases, including calf scours and paratuberculosis.

Bovine respiratory disease continues to be one of the most common endemic diseases of US cattle. The prevalence of BRD nationwide has been estimated at 12% and 5% in preweaned and weaned heifers, respectively (USDA, 2018). However, the burden of the disease far exceeds the reported prevalence due to the cross-sectional nature of prevalence as a measure of disease occurrence and the duration of BRD. As a result, cumulative incidence of BRD in preweaned calves, as the proportion affected from birth to weaning in a convenience sample of six California dairies from three major dairy regions, Northern CA, Northern San Joaquin Valley (NSJV) and Greater Southern CA (GSCA; includes Northern San Joaquin Valley and Southern CA), has been estimated at 22.8% (Dubrovsky et al., 2019a). Nationwide BRD accounted for 33.4% of all calf illnesses (Urie et al., 2020). Respiratory disease results in mortality and loss of productivity in the form of decreased milk production (Adams and Buczinski, 2016; Dunn et al., 2018a). An estimated 24% of preweaned and 59% of postweaning heifer mortality, respectively, are attributed to BRD, making it the second-most common cause of preweaned calf mortality and the top cause of mortality in postweaned calves (USDA, 2018).

BRD is a serious problem in the cattle rearing industry, and given its complex causal mechanism involving the host, pathogen and environment, it is a great disease candidate for a multimodal prevention approach. The goal of this chapter is to discuss the prevention of BRD as an example of bacterial disease prevention using the conceptual model of the disease triangle. The mechanisms related to the risk factors for each of the disease triangle components will be described. Finally, we will demonstrate a comprehensive and novel BRD risk assessment tool as a practical farm-ready application that can be used by producers, veterinarians, consultants and other calf raisers to identify and transect their operation-specific BRD risk factors.

1.2 Learning objectives

- Describe the basic physiology of the host, environment and pathogen relevant to the discussion of disease prevention, using the example of BRD;
- Review current literature and expert opinions describing the host, pathogen and environmental interactions relative to the BRD example;
- Describe recent advances in testing methods relevant to BRD;
- Identify interventions that can be used to interrupt the host–environment–pathogen continuum of disease; and
- Describe and provide detailed instructions for specific risk assessment and outcome assessment tools that can be implemented on the farm to facilitate disease prevention.

2 Pathogen host environment: an overview

Bacterial infection of the lower airways results from environmental conditions, pathogen virulence, a breakdown in host immunity or any combination of these factors. Different subsets of the BRD complex have been described relative to different cattle populations and risk factors. The clinical presentation of BRD in beef calves is commonly referred to as shipping fever, while in dairy calves, this syndrome is referred to as enzootic pneumonia (Callan and Garry, 2002); where applicable, these distinctions will be discussed.

2.1 Role of bacterial pathogens

Bacterial and viral pathogens are the primary pathogens implicated in the BRD complex, although other pathogens, such as parasites or fungi, and toxins can certainly result in respiratory disease in cattle. This section will focus on bacterial pathogens associated with BRD complex and factors that affect the role of these bacteria in the development of BRD.

The bacterial pathogens most commonly associated with BRD are *Mannheimia haemolytica*, *Pasteurella multocida*, *Histophilus somni* and *Mycoplasma bovis*. In addition, *Bibersteinia trehalosi* and *Trueperella pyogenes* are considered important bacterial pathogens associated with BRD; *T. pyogenes* is considered particularly important in chronic cases.

M. haemolytica, *P. multocida* and *H. somni* are all members of the Pasteurellaceae family and thus share some basic similarities. These organisms are gram-negative and thus possess an outer lipopolysaccharide (LPS), also known as endotoxin, which induces profound host inflammatory responses upon entering the tissues or circulation. All three organisms can also be found as commensal inhabitants of the upper respiratory tract in animals without signs of BRD (Allen et al., 1991; Gaeta et al., 2017).

2.1.1 Mannheimia haemolytica

M. haemolytica is often considered the predominant bacterial pathogen associated with BRD in feedlot cattle (Welsh et al., 2004; Klima et al., 2019) and among the top 3 in dairy calves (Francoz et al., 2015; Gaeta et al., 2017); recent reviews of this pathogen and its role in BRD have been published (Rice et al., 2007; Griffin et al., 2010; Confer and Ayalew, 2018). Briefly, there are at least 12 serotypes of this species and serotypes A1 and A2 are known to colonize the upper respiratory tract of cattle; type A2 is often isolated from clinically normal cattle, while type A1 predominates in the diseased cattle (Srikumaran et al., 1996; Rice et al., 2007). Other serotypes have also been isolated from lung tissues of cases of bovine pneumonia (Rice et al., 2007). *M. haemolytica* has several virulence factors that allow the organism to evade host responses, colonize host membranes and damage host tissues. These virulence factors include a capsule and adhesins for adherence and invasion; protective outer membrane proteins for host immune evasion; extracellular enzymes that interfere with the mucous layer; LPS, which leads to acute inflammation, edema, hemorrhage and hypoxemia; and leukotoxin (LKT), which is a pore-forming cytotoxin that leads to the lysis of neutrophils, macrophages and other white blood cells) (Rice et al., 2007; Gioia et al., 2006). Once this organism has gained access to the lower airway tissues, the combination of the effects of these virulence factors elaborates severe, acute clinical disease, often characterized by severe inflammation in the lungs and pleura (fibrinous pleuropneumonia, bronchopneumonia) that is often fatal. The clinical manifestation of the disease following *M. haemolytica* invasion into the lower airways is often called shipping fever, due to the association with a recent shipment of cattle to feed yards. Of course, the disease associated with this organism is not strictly restricted to beef cattle nor animals with a recent shipment history.

2.1.2 Pasteurella multocida

There are at least 16 serotypes of *P. multocida*; each possesses LPS and one of five known capsular types (A, B, D, E, F) but lacks the virulence factor LKT (Boyce and Adler, 2000). *P. multocida* type A:3 is the most common serotype isolated in cases of BRD (Dabo et al., 2008). Cases of BRD in which *P. multocida* predominates lower airway infection develop bronchopneumonia, and the clinical picture is usually less acute, with less dramatic signs of inflammation compared to infections where *M. haemoltyica* predominates. *P. multocida* is often identified in BRD of dairy calves (Francoz et al., 2015), and the clinical manifestation is termed enzootic calf pneumonia. Mortality rates for enzootic calf pneumonia have been estimated in the United States and Canada at 2.8% and 1.8%, respectively (Van Donkersgoed et al., 1993; Maunsell et al., 2011a). Surviving animals may have long-term damage, resulting in lower weight gain and decreased future milk production (Maunsell et al., 2011a).

2.1.3 Histophilus somni

In addition to LPS, *H. Somni* virulence factors include histamine production, biofilm formation, exopolysaccharide, outer membrane proteins and immunoglobulin-binding proteins. A fibrinopurlent bronchopneumonia results when *H. somni* infects the lower airway. The virulence factors, particularly an immunoglobulin-binding protein that affects endothelial cells, commonly allow the spread of the disease beyond the respiratory tract. In addition to fibrinopurulent bronchopneumonia, this organism has been associated with laryngitis, thrombotic meningoencephalitis, polyarthritis, pericarditis and myocarditis (Corbeil, 1996; Constable et al., 2016).

2.1.4 Mycoplasma spp.

M. bovis is the primary *Mycoplasma* species discussed in relation to BRD; however, other *Mycoplasma* have been associated with BRD in various roles. Different species of *Mycoplasma* impart different specific effects on the airway defenses. For example, *Mycoplasma dispar* causes ciliostasis and thus impairs upper airway clearance (Caswell, 2014). *M. bovis* colonizes and persists on mucosal surfaces and invades lung parenchyma, which induces a strong cellular response and elicits significant lung damage (Howard et al., 1987). Mycoplasmas can also impair bovine macrophage and neutrophil function (Howard and Taylor, 1983; Howard et al., 1987; Thomas et al., 1991). *Mycoplasma* organisms also predispose cattle to the invasion of the lower airways by other bacteria, and bacterial coinfections are common. Similarly to *H. somni*, *M. bovis* readily accesses the blood stream through the respiratory system; this can result in the disease of many organ systems; however, hematogenous spread leading to septic arthritis or otitis are some of the most common clinical manifestations of disease in calves. It should also be noted that *M. bovis* readily colonizes the eustachian tube, which serves as another route by which to cause otitis (Maunsell and Donovan, 2009; Maunsell et al., 2012). *M. bovis* can also cause mastitis; spreads from the mammary gland to other cows via milking processes or to calves through milk or colostrum are important routes of disease transmission (Maunsell et al., 2011b). *Mycoplasma* organisms are unique compared to the Pasteurellaceae previously discussed. They are surrounded by a trilayered membrane rather than a cell wall; this feature is particularly important when considering antibiotic therapy because this membrane structure makes *Mycoplasma* species inherently resistant to β-lactam type antimicrobials. Additionally, *M. bovis* has a family of variable surface lipoproteins, which undergo frequent phase and size change, exhibit strain variation, and can be selected by exposure to antibodies which contribute to immune evasion and presents a challenge to vaccine development (Lysnyansky et al., 2020; Le Grand et al., 1996; Beier et al., 1998; Buchenau et al., 2010; Maunsell et al., 2011b).

2.1.5 Other bacteria associated with BRD

Many other bacteria have been isolated from the lungs of pneumonic cattle in addition to the Pasteurellaceae and Mycoplasmas discussed earlier. The specific roles these other bacteria play in the BRD complex is less well defined (Constable et al., 2016). *Bibersteinia trehalose* is another bacterium in the Pasteurellaceae family, which also possesses an LKT (Murugananthan et al., 2018) and has been implicated in BRD, although the specific pathogenicity is debated (Adlam and Rutter, 1989; Kodjo et al., 1999; Katsuda et al., 2008; Hanthorn et al., 2014). *T. pyogenes* has been identified in many purulent infections in cattle; this organism may be isolated from chronic pneumonias or pulmonary abscesses (Hammer et al., 2004; Ribeiro et al., 2015).

2.1.6 Coinfections

The relationship between pathogens in the respiratory tract is also important to consider when discussing the pathogenicity of respiratory bacteria. Viruses and *Mycoplasma* have been the primary organisms implicated in coinfections with the respiratory Pasteurellaceae and predispose cattle to lower airway bacterial infection. It has long been established that viral respiratory infection predisposes cattle to bacterial bronchopneumonia (Jericho et al., 1986; Love et al., 2014, 2016a; Ng et al., 2015). The common viral pathogens implicated in BRD include bovine herpes virus 1 (BoHV1), parainfluenza type 3 (PI3), bovine respiratory syncytial virus (BRSV) and bovine viral diarrhea virus (BVD). Additionally, bovine coronavirus (BoCV) and Influenza D (IDV) have more recently been demonstrated to play a role in BRD (Hause et al., 2013; Ridpath et al., 2020). Viral pathogens are often implicated in the development of this disease complex as primary pathogens that weaken host defenses and allow the invasion of bacterial pathogens and opportunists into the lower airways. In these synergistic infections, viruses use many mechanisms to cause damage to host respiratory defenses and bacteria secondarily invade the lower airways. These mechanisms include structural changes such as injured respiratory epithelium, decreased mucociliary clearance and lung consolidation; biochemical dysfunctions such as decreased effectiveness or decreased secretion of antimicrobial immune compounds; and dysfunction of alveolar macrophages (Zachary, 2017). Ultimately, the resulting syndrome of BRD involves bacterial infection of the lower airways and the development of varying degrees of broncho, pleuro or interstitial pneumonia.

2.1.7 Respiratory microbial communities

More recent investigations into the relationships of different microbes in the respiratory tract of cattle have been studied using metagenomics. Historically, singular organisms, or the relationship between select organisms such as specific viruses and specific bacteria, have been studied in the pathogenesis of BRD. Recently, investigators have begun studying entire microbial populations in the upper and lower respiratory tract of cattle through the use of metagenomics. Not surprisingly, this work demonstrates diversity in the microbial populations of the bovine respiratory tract, and differences in these populations are associated with health and disease (Timsit et al., 2016a; Gaeta et al., 2017; Windeyer et al., 2017; Klima et al., 2019; McMullen et al., 2019). These microbial populations change over time (Timsit et al., 2016a; Stroebel et al., 2018) and vary between facilities (Stroebel et al., 2018). The presence of previously known pathogens, such as *M. bovis* (McMullen et al., 2019) and *M. haemolytica* (Klima et al., 2019), has not surprisingly been associated with BRD cases. Microbes not previously considered classic contributors to BRD, such as Bov rhinitis A virus, have also been associated with cases of BRD and require further investigation into their roles as pathogens (Ng et al., 2015; Zhang et al., 2019). The utility of studying entire populations of microbes in the respiratory tract is not simply to identify new or different pathogens but also to study the population dynamics of the respiratory tract microbiota that can reveal a deeper understanding of the interactions between the host, pathogen and environment. Metagenomic analysis has identified changes in the respiratory microbiome associated with environmental conditions (Nicola et al., 2017; Lima et al., 2019), diet (Hall et al., 2017), stressful events (Holman et al., 2017; Timsit et al., 2016b; Stroebel et al., 2018) and antibiotic treatment (Holman et al., 2018). The composition of the respiratory microbiome and those factors that may influence it are thus important considerations in the continuum of BRD causation and disease prevention.

2.2 Role of the host

Variability in host immunity is arguably just as important as the pathogens in the development of BRD. The relative susceptibility or resistance to BRD can be influenced by factors related to both acquired immunity and innate immunity.

2.2.1 Innate immunity

The bovine innate immune system of the respiratory tract is the collection of natural bodily defenses against pathogens, including physical and chemical barriers, mechanical effects, inherent cellular and chemical responses to foreign material or injury and the commensal microbes (Zachary, 2017).

The structure of the nose and paranasal sinuses acts to remove large particles from inhaled air (Zachary, 2017). Commensal microflora of the upper respiratory tract may also act to protect the airways from pathogens by a variety of proposed mechanisms (Ackermann et al., 2010). The large conducting airways are lined with pseudostratified, ciliated columnar epithelium and mucus-producing goblet cells. These cilia and mucus make up the mucocilliary escalator that functions to mechanically clear conducting airways of debris. Further down the respiratory tree are the transitional airways; this region is lined by cuboidal epithelium, fewer ciliated cells, Club cells (formerly Clara cells) that produce surfactant and protective proteins and neuroendocrine cells. Following the transitional airways is the primary gas exchange surface of the alveoli and alveolar ducts, which are lined with type I and type II alveolar cells (Zachary, 2017).

The lower airways were historically considered essentially sterile in healthy animals. However, recent investigations into the lower respiratory microbiome have revealed a variety of microbes present in the lower respiratory tract of cattle with and without signs of BRD (Johnston et al., 2017; Nicola et al., 2017; Timsit et al., 2018; Klima et al., 2019). These microbial populations are sampled from the large conducting airways (trachea and bronchi) and contain pooled samples from the terminal airways as well as these conducting airways.

Cattle are particularly predisposed to respiratory disease relative to other mammals for a variety of anatomic and physiologic reasons, including relatively low gas exchange capacity, greater anatomical compartmentalization, effect of gravity on ventral lung lobes, non-ciliated and ventral drainage of the pharyngeal tonsil, relatively low pulmonary compliance and relatively low numbers of pulmonary macrophages among others (Veit and Farrell, 1978; Kainer and Will, 1981). Many factors associated with environmental conditions in modern agriculture exacerbate these weaknesses in the bovine respiratory system, including the calf's ambient temperature, exposure to humidity, ammonia and dust among other factors, creating what is known as the animal's microenvironment. The role of these external factors is further discussed in Section 2.3, 'Role of the environment'.

2.2.2 Acquired immunity

Acquired immunity refers to the specific set of immune responses directed at specific types of pathogens and is also known as the adaptive immune system.

Calves acquire immunity in three general ways: passively from colostrum, actively through natural exposure to potential pathogens and deliberately through vaccination against target organisms.

2.2.2.1 Passive immunity

Calves are born agammaglobulinemic and thus rely on the colostral transfer of antibodies for humoral immunity. In addition, colostrum contains maternal leukocytes among other immune and growth factors. Calves rely on this maternally derived passive immunity until their own immune system matures. Inadequate colostrum intake in calves results in failure of transfer of passive immunity and thus failure of a primary immune defense in early life. The immunoglobulins are actively pinocytosed from the gut in the first 24 h after birth with a decreasing rate as time elapses and soon after first ingestion by the calf (Stott et al., 1979). Previous research identified serum immunoglobulin levels ≥1g/dL or serum total protein levels ≥ 5.2 g/dL as indicative of adequate transfer of colostral immunity. However, recent research suggests immunoglobulin concentrations greater than 2 g/dL or serum total protein >5.8 g/dL are associated with less risk for calfhood mortality (Chigerwe et al., 2015a). A similar cut-off (5.7 g/dL) was suggested to predict the risk of BRD (Windeyer et al., 2014). Feeding more high quality colostrum after birth, to allow for the associated increase in the transfer of passive immunity, and measurable increases in blood IgG concentrations in newborn calves continue to be a common goal on US dairies; the current research clearly shows that higher cut-offs would favor lower mortality and morbidity in calves (Urie et al., 2018a). New recommendations have been proposed to replace the current industry standard cut-off to assess the transfer of passive immunity for serum immunoglobulins of 10 g/L at 24-48 h after birth. The proposed consensus ranges for the transfer of passive immunity at the calf level include four serum immunoglobulin ranges ≥ 25.0 (excellent), 18.0-24.9 (good), 10.0-17.9 (fair) and < 10 g/L (poor) (Lombard et al., 2020). Lombard et al.'s (2020) consensus standards also call for a herd-level target of >40, 30, 20 and <10% of calves in the excellent, good, fair and poor transfer of passive immunity ranges, respectively, to reduce herd morbidity and mortality. Parallel ranges for serum total protein (g/dL) and serum Brix (%) are also provided for on-farm monitoring. The consensus ranges for serum total protein measurements are ≥6.2, 5.8-6.1, 5.1-5.7, <5.1 g/dL for the excellent, good, fair and poor transfer of passive immunity, respectively. The consensus ranges for serum Brix measurements are ≥ 9.4, 8.9-9.3, 8.1-8.8, <8.1% for the excellent, good, fair and poor transfer of passive immunity, respectively.

Due to the importance of the transfer of passive immunity to newborn calves, efforts to feed the optimum quantity and quality of colostrum are necessary to raise healthy calves. The quality of colostrum fed can be measured by immunoglobulin content and microbiological quality. Persistent udder infections were found to be associated with lower colostrum harvest volumes (Maunsell et al., 1998). As a result, in addition to harvesting colostrum from healthy cows, producers should expedite colostrum harvest soon after calving as one study estimated a 3.7% decrease in immunoglobulin content, with each hour delay subsequent to calving due to the initiation of milk secretion post-calving (Morin et al., 2010). Heat treatment of colostrum was found to reduce the total plate bacterial count and total coliform count, which may explain the reduced hazard for diarrhea, and any treatment from birth to weaning, in calves fed heat-treated colostrum compared to fresh colostrum (Godden et al., 2012). An association between the type of colostrum storage container and occurrence of respiratory disease on 100 dairies was identified by Maier et al. (2019a, 2020) with the use of bags to store colostrum compared to solid containers being associated with a reduction in cases; the latter association was significant in univariate models (single variable association with the outcome) only. However, the authors investigated potential confounders and effect modifiers, making this otherwise biologically significant association worthy of further investigation using an empirical approach, given the potential for reduction in odds of respiratory illness (Maier et al., 2020).

The ideal concentration of IgG in colostrum or ideal volume to be fed differs between studies and outcomes of interest (Windeyer et al., 2014; Chigerwe et al., 2015b; Urie et al., 2018b). Historically, 50 g/L IgG in colostrum was recommended as the bare minimum and calves were recommended to ingest at least 4 L to provide at least 200 g of IgG from initial colostrum feedings (Weaver et al., 2000; McGuirk and Collins, 2004; Chigerwe et al., 2008; Godden et al., 2009; Williams et al., 2014). Based on the data presented in the updated consensus by Lombard et al. (2020), this historical recommendation likely underestimates the colostrum quality and total mass of IgG needed to ingest to achieve excellent passive immunity (defined as $\geq$ 25 g/dL serum IgG). In calves with excellent passive immunity, the mean IgG ingested was 286.7 g in one feeding or 421.2 g in calves fed multiple feedings; this mass of IgG was delivered in an average of 3.3 L and 5.3 L in the first 24 h, respectively (Lombard et al., 2020).

Producers should also consider actively monitoring the transfer of passive immunity in newborn calves by measuring serum total proteins from a sample of calves as a proxy for blood immunoglobulin levels 24 h post colostrum feeding (Pithua et al., 2013). In addition, routine assessment of a dairy's colostrum management should include staff training on the best practices for colostrum harvest, quality in terms of immunoglobulin content and bacterial concentration, heat treatment, storage conditions including container type and volume and time from birth to feeding to calves. Special attention to these practices is needed when feeding pooled colostrum by training calf caretakers and dairy staff involved in all aspects of colostrum management to achieve higher immunoglobulin concentrations in the pooled colostrum fed to calves. Specialized training interventions have been shown to create measurable improvements in calf serum total protein (Williams et al., 2014).

2.2.2.2 Vaccines

Vaccination is a popular method for improving host defenses, via acquired immunity, against specific respiratory pathogens (Windeyer et al., 2012; Chamorro and Palomares, 2020). The vaccines currently marketed in the United States for the prevention of BRD include *M. haemolytica*, *P. multocida* and *H. somni* and the viral pathogens BVD, BRSV, IBR and PI3. Although the viral disease is not the focus of this chapter, the importance of viral vaccination for the prevention of BRD is an important consideration due to the synergistic role viral pathogens demonstrate with bacterial pathogens in the development of BRD; a full discussion on viral vaccination for BRD is beyond the scope of this chapter. Broadly speaking, modified live viral (MLV) respiratory vaccines induce both cell-mediated and humoral immune responses in the host, while killed viral vaccines produce humoral immunity but less significant cell-mediated immunity (Platt et al., 2009; Woolums et al., 2013; Chamorro and Palomares, 2020). These different host responses allow fewer doses to achieve the desired protection and longer-lasting immunity with MLV vaccines, whereas a minimum of two doses, spaced approximately 21 days apart to allow induction of an anamnestic response, are typically required for killed viral vaccines (Chamorro and Palomares, 2020; Ridpath et al., 2020). Modified live viral vaccination, however, can present a risk in breeding animals due to effects on the early conceptus or semen production, and label instructions relating to breeding safety should be followed when creating vaccination schedules, particularly with MLV vaccines (Grooms et al., 1998; Nandi et al., 2009; Palomares et al., 2013). Early vaccination of calves using injectable MLV vaccines, or intranasal MLV vaccines, in the face of maternal antibodies is a strategy that may provide early protection and/or immune priming for some viral pathogens; the calf may or may not seroconvert, and cell-mediated immunity appears to be an important factor in this strategy, particularly with intranasal vaccines (Stevens et al., 2011; Gerber et al., 1978; Vangeel et al., 2009; Woolums et al., 2013; Mahan et al., 2016; Windeyer and Gamsjäger, 2019; Chamorro and Palomares, 2020). However, the effect of vaccination in the face of maternal antibodies depends on several factors, including the pathogen and concentration of maternal antibodies, and significant maternal antibody inference has been identified in some studies, and there is no consensus on the universal utility of early calfhood vaccination in the face of maternal antibodies (Downey et al., 2013; Ellis et al., 2001, 2014; Windeyer and Gamsjäger, 2019; Chamorro and Palomares, 2020).

Vaccination of cattle against viral pathogens associated with BRD is often considered a cornerstone of BRD prevention and is commonly used (USDA, 2010; Waldner et al., 2019). Vaccination against bacterial pathogens of BRD is less commonly employed than vaccination for viral pathogens (Karle et al., 2019). Multiple different strategies have been used to create vaccines for bacterial BRD pathogens, primarily the Pasteurellaceae (McGill and Sacco, 2020). Surface antigens of the Pasteurellaceae, and LKT in *M. haemolytica* specifically, have been a focus of vaccine development (Confer and Ayalew, 2018). A 2012 meta-analysis investigated the effect of vaccination for *M. haemolytica*, *P. multocida* and *H. somni* on morbidity and mortality in feedlot cattle; this analysis reports conflicting results between studies, with an overall assessment that vaccination for *M. haemolytica* and *P. multocida* is potentially beneficial, while no evidence for the benefit of *H. somni* vaccination was identified (Larson and Step, 2012). In *M. bovis*, the variability in *M. bovis* surface proteins may contribute to immune evasion and presents a challenge to vaccine development (Le Grand et al., 1996). Previous vaccination attempts for *M. bovis* have been unsuccessful; in a randomized, placebo-controlled, double-blinded field trial, vaccination for *M. bovis* had no effect on treatment for *M. bovis*-associated disease, nasal colonization with *M. bovis*, serum antibody titers for *M. bovis* or mortality (Maunsell et al., 2009).

The reduction of BRD mortality through the use of viral vaccines, bacterial vaccination or a combination is not universally successful (Larson and Step, 2012; Theurer et al., 2015; Chamorro and Palomares, 2020). Studies of vaccine efficacy often appear promising in controlled clinical trials; however, findings from large field studies often differ. These discrepancies are not only due to variability in study design (discussed later) but also likely because vaccination inherently affects one or two narrow aspects of the disease triangle. Specifically, the host's adaptive immunity is affected by vaccination, and arguably the environment may be affected through the creation of herd immunity, thus decreasing the environmental load of a given pathogen in the herd. It is important to remember that other host attributes, environmental conditions and pathogen factors can all influence the occurrence of BRD. For example, if cattle are stressed (Blecha et al., 1984) or malnourished (Griebel et al., 1987; Galyean et al., 1997), or if vaccination is not timed properly relative to disease risk (Lippolis et al., 2016; Richeson and Falkner, 2020), the immune response may be altered and thus insufficient to protect against clinical manifestations of BRD. It is imperative to remember that although specific pathogen vaccination is an attractive prevention strategy for BRD, this intervention alone may be insufficient to provide rewarding results in disease prevention. The authors recommend the thoughtful use of vaccination, on a farm-by-farm basis, based on knowledge of farm-specific disease and risk periods, in combination with other management strategies as discussed in this chapter, to modify all three arms of the disease triangle to provide the most impactful BRD disease prevention.

2.2.2.2.1 Veterinary vaccine research design and implementation

Vaccine trials play an important role in validating findings from research at the development phase. Specifically, vaccine efficacy can be estimated using controlled experiments that may include challenges by a live pathogen(s). There is an important distinction between the reported efficacy of a vaccine and its effectiveness in a field setting. The nomenclature of vaccine efficacy refers to findings produced by empirical studies such as double-blind block randomized trials, which continue to be the gold standard of cause and effect. However, the nomenclature vaccine effectiveness refers to findings based on studies conducted on a wide range of management practices to provide meaningful vaccine effectiveness estimates that can then apply to a wide range of production systems. Measurement of the host immune system reaction, inflammatory response, shedding (if applicable), feed intake and productivity measures, along with the economic impact, are also needed to properly assess the effectiveness of a vaccine.

Among the challenges encountered when conducting vaccine research is the availability of a natural challenge to accurately represent field conditions. Variability in field conditions include differences in natural challenge dose, intensity and strain variety. Hence, documenting the natural challenge, both duration and intensity, through continuous surveillance in the study population during vaccine trials is necessary to interpret trial findings. The inclusion of study subjects that encompass different management practices, such as both calf ranches and dairies, large and small herds, animals raised under different environmental conditions, different seasons, regions and climates, is a way to increase the external validity when estimating the effectiveness of vaccines.

Nevertheless, field conditions may produce immunization rates that differ from efficacy studies that precede vaccine effectiveness research. Other reasons for such differences include variability in vaccine production, cold chain – if applicable, dosing and administration, vaccination age and host immune status, which can be subject to nutrition, housing or other factors of the environment arm of the triangle. Capturing such variability in the field provides more realistic estimates that capture the true vaccine effectiveness as opposed to efficacy. As a result, to estimate vaccine effectiveness, alternative study designs should be pursued including retrospective case-control studies or prospective cohort studies (Knight-Jones et al., 2014; Weinberg and Szilagyi, 2010). Such observational study designs, based on several herds with different natural challenge exposures, allow for the measurement of the true effectiveness of a vaccine (Gardner et al., 2007; Perrett et al., 2018).

Finally, structuring the hypotheses remains a key factor in the success of research on vaccine effectiveness. Veterinary vaccine trials should explore herd immunity rather than simple measures of the percent immunized, disease hazard and occurrence in vaccinates compared to controls. Using herd immunity, veterinary vaccine trials should estimate the percent of the population that is needed to be immunized in order to eradicate the disease in a specific herd or, if the latter is not possible, explore the duration of vaccine immunity and the necessary repeated immunization frequency and rate based on waning immunity rate (Woolhouse et al., 1997).

2.2.3 Influences on host immunity

Both the innate and adaptive immune systems can be influenced by many host factors; common examples of such factors include stress, nutrition and genetics.

2.2.3.1 Stress

Many factors may cause stress in cattle, and the underlying cause of the stress may also induce direct damage to the host's immune defenses. Depending on the cause and chronicity of the stressor, the endogenous stress responses of the animal can have differing effects. The endogenous stress response involves variable stimulation of the sympathetic nervous system (SNS) and hypothalamic pituitary axis (HPA), leading to the release of catecholamines, cortisol and other endogenous stress responses. These responses have been demonstrated to alter host immune responses in a variety of ways ranging from beneficial priming of immune responses in acute stress to detrimental lack of response or even self-destructive autoimmune responses in states of chronic stress exposure (Mansfield et al., 2008; Carroll and Forsberg, 2007; Ackermann et al., 2010). Stress in cattle induced increased replication and spread of bacteria inhabiting the upper respiratory tract, including P. multocida, M. haemolytica and H. somni (Ackermann et al., 2010). Additionally, in vitro research has demonstrated direct interactions between pathogenic bacteria and stress hormones including epinephrine and norepinephrine that provide evidence for increased pathogenicity of bacteria in the presence of mammalian stress hormones (Lyte, 2004). The bovine respiratory tract contains abundant blood flow and noradrenergic innervation, so it is conceivable that stress not only alters host immune responses but may also directly alter pathogen virulence in the respiratory tract (Lyte, 2014a).

Heat stress is an extremely common environmental stressor, and, as global warming takes place, heat stress may become increasingly problematic in cattle production systems. In addition to the stress response associated with HPA stimulation, heat stress is associated with GI and systemic inflammation (Kim et al., 2018; Koch et al., 2019). The proposed mechanism for the inflammation identified in heat stress is that heat (or a high-temperature humidity index) exposure triggers increases in peripheral blood flow via vasodilation as the body attempts thermoregulation. This peripheral vasodilation occurs at the expense of internal vasoconstriction to maintain blood pressure and vital organ perfusion. The GI tract may become hypoperfused during heat stress, and intestinal ischemia leads to dysfunction of intestinal epithelial cells, as evidenced by changes in intestinal villous integrity (Pearce et al., 2013a,b) and dysfunction of intestinal epithelial cell tight junctions (Koch et al., 2019). The delicate barrier between the enteric lumen and the bloodstream is thus compromised, and local and systemic immune mediators react as LPS, bacteria or other luminal toxins translocate to portal circulation. This inflammatory response associated with heat stress uses energy (Baumgard et al., 2011). One-month-old calves' thermoneutral environment is estimated to range in temperature from 10°C to 25°C, with a maximum limit of 30°C for maximum acclimation, which is narrower than that for adult cattle (Wathes et al., 1983; Davis and Drackley, 1998). Cattle that are exposed to environments beyond the thermoneutral zone use energy to release heat from the body through rapid breathing, resulting in reduced feed intake and reduced immune function due to the calf's energy resources being diverted to address heat stress (Lammers et al., 1996; Mitlöhner et al., 2002). Lower alveolar macrophage counts and higher monocyte counts were associated with temperature and humidity in calves, further suggesting that heat stress may suppress a calf's respiratory immunity (Bertagnon et al., 2011; Roland et al., 2016). The combined effects of inflammation, stress response and decreased feed intake all probably play a role in predisposing cattle with heat stress to BRD.

2.2.3.2 Nutrition

Proper nutrition is essential to the maintenance of homeostasis, including immune function (Griebel et al., 1987). The complex life cycle of dairy cattle requires that each stage of life is supported by a plane of nutrition that matches the demands for that specific life stage in a given environment.

Feeding calves less than 2.84 L (3 US quarts) daily was associated with the increased respiratory disease with varying degrees depending on the calves' breed, with the greatest negative impact seen in Holsteins compared to Jerseys (Maier et al., 2019a). In contrast, the latter study also found that feeding more than 5.68 L (6 US quarts) per day benefited Jersey calves more than Holsteins, documenting further benefits to reducing morbidity due to respiratory diseases by feeding more milk to unweaned calves, and while such benefits may be observed in Jersey calves by exceeding 5.68 L, more research is required to identify such cut-off for Holsteins (Maier et al., 2019a). Similar findings were reported in a longitudinal study that estimated a 92% reduction in the hazard of respiratory disease when calves were fed 4.25–5.68 L of milk daily (Dubrovsky et al., 2019a). Lower volumes of daily milk fed, and hence less energy, may not satisfy the calf's metabolic maintenance, growth and immune system requirements (Drackley, 2008). The quality of milk fed may also impact the health of calves. Calves fed strictly milk replacer had a higher risk of morbidity compared to calves fed pasteurized waste milk (Godden et al., 2005). Milk replacer may not be as nutrient rich to support a calf's growth and immune system requirements, compared to cow's milk in either saleable or waste milk form. In addition, calves fed saleable and/or hospital milk had lower respiratory disease risk compared to those fed the same sources when supplemented with milk replacer, further emphasizing the impact of the nutrient content of a calf's diet on its risk for respiratory disease (Dubrovsky et al., 2019a) as was feeding saleable milk to calves associated with lower prevalence and of respiratory disease in calves compared to non-saleable milk (Maier et al., 2019a). The quality of milk fed to calves from a bacteriological aspect may also impact the occurrence of the respiratory disease; Maier et al. (2019a) estimated from a cross-sectional study of 100 dairies that calves fed non-pasteurized milk had higher odds of respiratory disease compared to those fed pasteurized milk. The same study also identified an association of feeding order by age where feeding older calves first was associated with a higher prevalence of respiratory disease. Producers and herd managers should also consider the necessary changes in calf diets toward weaning age to avoid complications due to improper rumen development. Among the changes that should be considered are a gradual decrease in the volume of milk diet fed, offering properly formulated solid diets and extending the weaning age to allow for rumen development.

2.2.3.3 Genetics

Recent genomic investigations have identified genetic differences associated with susceptibility or resistance to BRD (Van Eenennaam et al., 2014). Quantitative trait locus mapping has been used to find regions linked to BRD susceptibility (Casas et al., 2011; Glass et al., 2012); gene set enrichment analysis has been used to find upstream regulators of BRD susceptibility in beef and dairy cattle (Neupane et al., 2018). Although this investigation into genetic selection for BRD resistance is relatively new, genomic testing for wellness traits, including respiratory disease, have recently become commercially available. The percent of the disease explained by the genotype varies between studies and geographic location (Glass et al., 2012; Hoff et al., 2019), and heritability estimates for BRD resistance are low (Snowder et al., 2006). This area of research is still developing, and the herd effects of genetic selection for BRD resistance on BRD morbidity or mortality are still unknown. The continued investigation into the genetic backbone of immunity will hopefully provide selection criteria, aid in vaccine development and help us better understand host responses to the pathogen and environmental pressures.

2.2.3.4 'Upregulators' of immunity

Specific pharmacologic interventions intended to improve the innate immune system have recently been developed and are an active area of study. In contrast to vaccination, which primes specific immune responses characterized by antigen-specific humoral responses, this new strategy seeks to increase the general responsiveness of the innate immune system and is termed 'trained immunity'. One such product is a DNA-based immunostimulant, which apparently activates the cGAS-STING pathway; this is a mechanism to enhance host detection and response to cytosolic foreign DNA (Ilg, 2017). Another product marketed for innate immune stimulation is derived from the cell wall of the non-pathogenic *Mycobacterium phlei* (Nosky et al., 2017; Romanowski et al., 2017). The exact mechanism of immune stimulation of this product has not been clearly described but is postulated to involve mycobacterial compounds trehalose 6,6'-dimycolate (TDM) and muramyl dipeptite (MDM). These compounds are reported to enhance phagocytosis and antibody-mediated cytotoxicity and enhance lymphocyte activation (O'Reilly, 1992; Traub et al., 2006); however, their overall effectiveness in the prevention of disease in herds is unknown.

2.3 Role of the environment

Environmental factors can have drastic effects on the respiratory system; such factors include heat and cold stress; housing and hygiene practices that may expose calves to high pathogen loads; poor air quality, particularly with air pollutants such as dust, ammonia or other pollutants; and additional physical stressors such as with shipping or processing conditions. The environment can impart psychological stressors as well, such as with handling, shipping and changes in social structure during commingling. The environment can also affect changes in the commensal respiratory microbiota that influence host susceptibility to BRD (Lima et al., 2016; Nicola et al., 2017; Timsit et al., 2018). Many different types of environmental factors can affect the systemic immune defenses, especially those related to BRD.

A common environmental factor that influences BRD is heat. As discussed earlier, under Section 2.2.3, 'Influences on host immunity', heat stress may interfere with host defenses to BRD. There are several factors within the environment that may alter heat or heat stress exposure. For example, heat stress may be magnified within a calf's hutch compared to the ambient area, increasing the risk of respiratory disease (Louie et al., 2018). Specifically, Louie et al. (2018) estimated higher odds of BRD in calves as the daily maximum temperature increases during summer months. Moreover, the magnitude of the association was greater when the daily maximum temperature was recorded within the hutch compared to the ambient area, further demonstrating a difference in the calf environment within (microenvironment) compared to outside (macroenvironment) the hutch. Differences between calves' micro- and macroenvironment have been observed elsewhere (Macaulay et al., 1995; Carter et al., 2014). Such temperature and humidity changes impact calves' risk for respiratory disease year-round but with seasonal variation. Dubrovsky et al. (2019a) reported that the risk of respiratory disease in calves was the greatest in the spring followed by summer, compared to winter, with no difference between the fall and winter seasons. A cross-sectional study that captured greater geographic variability and a wider range of climate zones in California showed that fall was associated with the greatest odds of respiratory disease compared to spring, with the remaining seasons not different from spring. Both studies point to the transitioning between hot and cold seasons, which raises the question of whether fluctuations in temperature over a short period of time, including overnight, may be associated with a greater risk of respiratory disease in calves, at least in dry climates. Abrupt temperature fluctuations may also impact a calf's respiratory system's ability to clear *M. haemolytica* (Jones, 1987). Further research is needed on abrupt temperature and/or humidity changes as risk factors for respiratory disease in calves.

In the United States, approximately 70% of preweaned dairy calves are raised individually in hutches (USDA, 2016), which may vary in design, material, size and amount of protection from heat and cold stress. In addition to the specific environment within a calf's enclosure, the material such housing is made off further dictates the calf's exposure to environmental stress. Karle et al. (2019) reported that Northern California calves housed in hutches with metal components had significantly higher BRD prevalence compared to calves housed in wooden hutches (15.3% versus 3.2%). Maier et al. (2019a) found that calves housed in hutches constructed with metal components (walls or roof) had 11.2 times the odds of BRD compared to those housed in wooden hutches. A longitudinal study of more than 11 000 calves followed from birth to weaning similarly estimated a greater risk for BRD in calves housed in hutches with metal components compared to those housed in wooden hutches (Dubrovsky et al., 2019a). Raising calves in hutches with a metal roof or wall may be associated with greater BRD occurrence since metal is a better heat conductor compared to wood; this may result in greater temperatures for longer durations in hot summers. Furthermore, shade cloth or similar external shade structure above the calf-raising area has been associated with a reduced calf respiratory disease (Maier et al., 2019a). Modifying the environment to reduce the risk of bacterial infections can be beneficial for preventing BRD in calves. The impact of heat stress on the bovine respiratory system, and its role in respiratory diseases, can be further amplified by environmental conditions that modify the animal's microenvironment, creating additional adverse conditions that predispose to respiratory disease. For example, exposure to recycled lagoon water, used to flush fecal slurry below raised hutches with slatted wooden floors, was found to be associated with increased respiratory disease (Maier et al., 2019a). Other environmental factors that impact calves' risk for respiratory illness include air quality. A longitudinal study following 11 300 calves on five dairies in California identified that the use of water trucks for dust control resulted in a greater incidence of calf respiratory disease; while such results may seem contradictory, the study investigators noted that trucks traveled on dusty roads around the calf-raising area at speeds that often generated more dust (Dubrovsky et al., 2019b). In the same study, the owner reported that dust in the calf-raising area was also associated with an increased risk of respiratory disease, with dust around calves seen regularly being associated with the greatest risk of respiratory illness (Dubrovsky et al., 2019a). Dust can be a source of inhaled endotoxin and has been shown to cause coughing, fever, decreased feed intake, leukocytosis and mild interstitial pneumonia in sheep (Purdy et al., 2002). In contrast, magnesium chloride has been used in equestrian arenas to control dust and has been used on some dairies for the same reason. Indoor housing systems similarly can alter the calf's environment due to either improper ventilation or heating systems, leading to higher concentrations of dust and/or ammonia that may impact the calf's respiratory system (Nordlund and Halbach, 2019). Other studies reported an association between filtered air from indoor barns and a lower incidence of respiratory disease in calves (Pritchard et al., 1981; Hillman et al., 2010).

Environmental exposure to pathogens can also influence the occurrence of BRD. Starting at the time of birth, the calf is exposed to maternal microbiota; the composition of maternal vaginal microbiota appears to correspond to the calf's upper respiratory microbiota and alter the calf's risk of BRD (Lima et al., 2019). Bedding hygiene in the maternity pen also alters the risk of BRD in calves. Dubrovsky et al. (2019a) reported a 2% drop in the hazard of BRD with every additional change in maternity pen bedding per month. The latter study followed calves born in individual or group pens (one dairy), group calving pens (one dairy), individual calving pens (three dairies) or pasture (one dairy) and also estimated a decrease in the risk of BRD mortality when comparing maternity pens with bedding changed four to nine times compared to less than four times per month. An increase in the frequency of maternity bedding changes is likely to reduce the concentration of pathogens a calf is exposed to at or soon after birth. A similar association can be expected when contrasting individual to group maternity pens. Svensson et al. (2003) reported a 46% reduction in odds of BRD in calves born in individual pens compared to group maternity pens. Furthermore, bedding material used can also have an impact on the occurrence of respiratory disease in calves. Calves born in maternity pens bedded with dried manure or sand had a lower prevalence of respiratory disease compared to those born on dirt or plant materials, including rice hulls, almond shells, straw, wood shavings or wood chips (Maier et al., 2020).

The respiratory microbiota of cattle can influence that of other in-contact animals, and the composition of these microbes appears to change relative to many other factors, including the farm of origin (Nicola et al., 2017), diet (Hall et al., 2017), stress (Holman et al., 2017; Timsit et al., 2016b; Stroebel et al., 2018) and antibiotic treatment (Holman et al., 2018). For example, the nasopharyngeal microbiota of feedlot cattle changes between arrival and after 60 days on premise (Holman et al., 2015). In another study, antibiotic-resistant clones of *P. multocida* appeared to spread to arriving calves after arrival to the feedlot (Guo et al., 2020). The animal-to- animal contact, which varies with housing setup, is likely a factor that influences the transmission of microbes and alterations in the respiratory microbiota; placement of hutches in close proximity that allowed calf-to-calf contact was associated with an increase in BRD in calves greater than 75 days old when compared to calves of the same age group without calf-to-calf contact (Maier et al., 2019a). Furthermore, within cohorts of calves with calf-to-calf contact, older calves consistently had greater odds of developing BRD when compared to younger calves. Further research is needed to investigate the role of animal contact (within or between ages) and respiratory disease, specifically as it relates to changes in the upper and lower respiratory microbiota.

2.4 Other disease syndromes

Scours and respiratory disease continue to be the most common causes of preweaned dairy calf morbidity and mortality (USDA, 2018). Despite their distinct pathology and clinical presentations, these two syndromes are good examples of different diseases that share many causal mechanisms of the host, pathogen and environment. Host immunity is of paramount importance, particularly in young calves that are more susceptible to infectious diseases until the immune system matures and can better react to new exposures; during early life the calf relies heavily on the transfer of passive immunity through appropriate colostrum intake. Inappropriate colostrum management can predispose calves to both respiratory and enteric diseases. The environment in which the calf is born and reared in provides variable protection from the elements; environmental temperature extremes can stress the calf and impact calf immunity. The hygiene of the calfhood environment also determines exposure to both respiratory and enteric pathogens. These are just a few examples of how both of these very common calfhood disease syndromes ultimately share many of the same risk factors classified under the host, pathogen and environment. Several of the prevention and control interventions discussed in this chapter and proposed as part of the BRD risk assessment tool described later in the chapter may similarly interrupt the causal web for other diseases.

3 Disease detection

An important part of disease prevention is appropriate disease detection. The clinical signs of respiratory disease can vary in type and severity, and some affected animals may show no obvious outward signs at all (Buczinski et al., 2014; Ollivett et al., 2015). Acutely affected animals may show different signs than chronically affected animals. These features of the disease have led to a variety of different techniques used to aid in the diagnosis of BRD. There are many tools that can be used to detect or further characterize cases of respiratory disease and the pathogens associated with respiratory disease.

3.1 Disease detection techniques

3.1.1 BRD scoring systems

Availability of scoring systems for health outcomes such as BRD allows for rapid and low-cost detection of cases. BRD scoring systems work by the assessment of clinical signs, or risk factors, and produce a score for each animal, which is compared to a cut-off value. A score that equals or exceeds a given cut-off classifies the patient as BRD score positive. Users can then interpret such score positive status as the presence of the condition or disease within a certain level of sensitivity and specificity provided by the specific scoring system used. Scoring systems have been previously proposed for respiratory disease in dairy and beef cattle (Thomas et al., 1977; Panciera and Confer, 2010). Two systems have been used in recent studies for BRD in preweaned dairy calves: the Wisconsin (WI) BRD scoring system developed at the University of Wisconsin; and the California (CA) BRD scoring system developed at the University of California Davis. The latter was based on a pair-matched case–control study of 2,030 preweaned calves (Love et al., 2014, 2016a). Both systems assess clinical signs associated with respiratory disease including nasal discharge, ocular discharge, cough, fever and ear disposition; however, the CA system assesses breathing quality as a sixth clinical outcome. Breathing quality is assessed for a calf by comparing calf respiratory rate and difficulty to the other calves in the same herd at the same time of scoring; this adjusts for the effect of ambient environmental conditions such as heat stress. The WI system scores each clinical sign on a gradient of severity, while the CA system dichotomizes each into absent or present. The WI system scores each clinical sign in single-unit increments from 0 (normal) to 3, and the CA system scores vary by the clinical sign as a function of the strength of the association between the respective clinical sign and respiratory disease as measured in a pair-matched case–control study (Love et al., 2014). Specifically, the CA system scores nasal discharge, occular discharge, cough, fever, ear disposition and breathing quality as 4, 2, 2, 2, 5 and 2 points, respectively (Figs 2A and 2B). Both systems assign a calf BRD score positive status at scores ≥ 5. In addition, the WI system assesses both spontaneous and induced cough, while the CA system assesses spontaneous cough only, which is a tradeoff to avoid calf handling to maintain biosecurity between calves. The CA scoring systems

Figure 2A The California bovine respiratory scoring system. (Image reprinted by permission from Dr. Sharif S. Aly.)

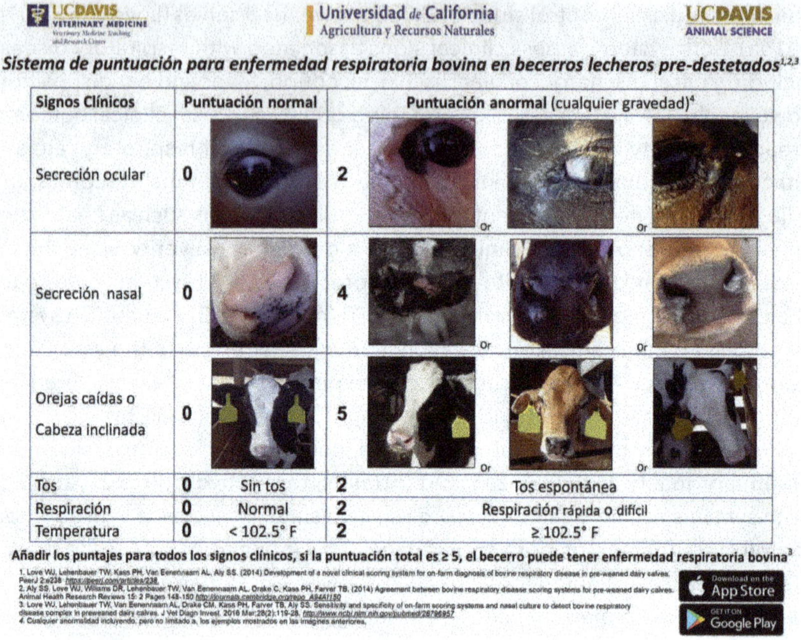

Figure 2B Spanish translation of the California bovine respiratory scoring system. (Image reprinted by permission from Dr. Sharif S. Aly).

call for less calf handling since only calves observed with a total score of 4 would require a rectal temperature to ascertain whether its final score exceeds the cut-off of 5 points. In a separate nested case–control study of 536 calves on three dairies and two calf ranches in California, both the CA and WI scoring systems had similar sensitivity when screening calf herds for BRD (46.8% versus 46.0%, respectively) and a similar diagnostic sensitivity when detecting a case (72.6% versus 71.1%, respectively); however, the specificity of the WI system was greater than that of the CA system (91.2% versus 87.4%). Despite the difference in specificity between both systems being statistically significant, it's unlikely to be biologically important; however, users should weigh such a difference against ease of use. The CA scoring system has the advantage of simple dichotomous classification for each clinical sign and the fewer calf handling requirements (no test for induction of cough and rectal temperature needed only on calves with a score of 4), which improve biosecurity.

Weaned and growing dairy youngstock are also affected by respiratory disease. In fact, respiratory disease in the United States is the greatest cause of morbidity and mortality in dairy youngstock postweaning (USDA, 2018). A scoring system recently developed at the University of California Davis for weaned and growing dairy youngstock was based on a prevalent case–control study on five dairies (Fig. 3). The development of the CA BRD scoring system for post-weaned calves followed the same approach as the preweaned scoring system with cross-validation for estimating its accuracy; the latter is a technique where the dataset is split such that observations are used only once for development or validation but not for both (Maier et al., 2019b).

3.1.2 Thoracic ultrasound

Diagnostic imaging adds to the clinical exam and can be particularly useful to confirm suspected cases, characterize the distribution or severity of the disease and detect sub-clinically affected animals (Ollivett et al., 2015). Thoracic ultrasound is a popular imaging modality in calves due to the practical on-farm utility of this technique; the linear probe typically used for bovine reproductive work can be positioned between ribs to image much of the lung surface (Ollivett and Buczinski, 2016). The technique is relatively simple to perform after training and requires only basic restraint, a coupling agent such as alcohol or gel, occasionally clipping if the hair is thick or dirty, and an ultrasound unit. For on-farm use, typically the linear rectal probe is used; however, other

California bovine respiratory disease scoring system for post-weaned dairy calves

Clinical sign	Score if normal		Score if abnormal (any severity)[1]
Sunken eyes	0	4	Or
Low body condition	0	5	Or
Cough	0	No cough	2	Spontaneous cough
Breathing	0	Normal	1	Rapid or difficult breathing
Diurnal temp fluctuation	0	≤27° F (≤15° C)	1	> 27° F (>15° C)

With diurnal temperature data: calf is score positive[2] if total score ≥ 2		Without diurnal temperature data: calf is score positive[2] if total score ≥ 1

Confirmatory step for score positive[3]	Do not treat	Treat
Rectal temperature	< 102.5° F (< 39.2° C)	≥ 102.5° F (≥ 39.2° C)

1. Any abnormality including, but not limited to, the examples shown in the above pictures. For body condition, a score of ≤ 2 on a scale of 5 would be considered abnormal.
2. With diurnal temperature fluctuation: screening sensitivity 77%, diagnostic sensitivity 100%, specificity 62%, without diurnal temperature fluctuation: screening sensitivity 84%, diagnostic sensitivity 100%, specificity 46%
3. With diurnal temperature fluctuation: confirmatory step improves specificity to 77% (screening sensitivity 65%, diagnostic sensitivity 77%), without diurnal temperature fluctuation: confirmatory step improves specificity to 63% (screening sensitivity 71%, diagnostic sensitivity 77%)

Figure 3 California bovine respiratory scoring system for post-weaned dairy calves (Used by permission from *Journal of Dairy Science*; vol. 102:7329–7344; doi.org/10.3168/jds.2018-15474).

ultrasound probes can likewise obtain useful images of the lung surface. The depth and focal distance can be adjusted to improve image quality at the lung surface based on the patient's chest wall thickness, probe type used and depth of consolidated lesions. Healthy, aerated lung creates an echogenic line at the lung surface followed by a reverberation artifact that is readily identifiable. Pneumonia causes inflammation in the lungs that leads to fluid accumulation, cellular infiltrates and debris that alter the gas-containing spaces. Areas of diseased lung thus lose the characteristic gas echogenicity and reverberation on ultrasound and appear as soft tissue echogenicity similar to the liver; large areas of consolidation are sometimes referred to as 'hepatized lung' (Reef et al., 1991; Ollivett et al., 2013). Pleural irregularities are visualized on ultrasound as irregularities in the echogenic surface of the lung and are termed "comet tails'; as these areas converge from individual lesions to larger continuous irregularities, they are termed 'B-lines' (Adams and Buczinski, 2016). Pleural effusion is readily apparent on ultrasound as hypoechoic fluid outside the lungs, and abscesses appear as mixed echogenicity lesions typically contained in a more echogenic capsule. Different scoring systems have been developed to classify ultrasonographic lung lesions and have been used in both research and clinical settings, on individuals and herds, to characterize lung disease and identify sub-clinically affected animals (Adams and Buczinski, 2016; Ollivett and Buczinski, 2016). The findings of thoracic ultrasound can be useful to prognosticate, assign treatment, assess response to treatment and screen groups of animals. Systematic screening for BRD in specific animal groups, using an ultrasound scoring system, may be useful for veterinarians to identify when populations at risk for BRD first start to have evidence of the disease and creates an objective measure on which to form animal health goals and assess response to preventative interventions (Calf Health Module - The Dairyland Initiative; Ollivett and Buczinski, 2016; Dunn et al., 2018b). Other respiratory imaging techniques that are occasionally used clinically, or in specific research settings, but are not usually implemented on a herd basis include airway endoscopy, radiographs, CT/MRI and fluoroscopy. These alternative advanced imaging techniques can be useful in select cases or in specific research investigations that require detailed thoracic imaging. These techniques are not employed in herd screening methods due to the cost, limited availability of equipment and impractical nature for imaging large numbers of animals.

3.1.3 Auscultation

Thoracic auscultation using a standard stethoscope can be performed to assess for abnormal lung sounds, which may support a diagnosis of BRD; however, thoracic auscultation alone has very poor sensitivity for detecting lung lesions (Buczinski et al., 2014). A specialized lung auscultation device has been marketed to aid in the detection of BRD in feedlot cattle. This computer-assisted stethoscope records sounds during thoracic auscultation and applies a computer algorithm to generate a BRD severity grade, ranging from unaffected to chronically affected, to animals showing signs of illness. In a study comparing standard thoracic auscultation to this device, the agreement between the two methods had a kappa value of 0.77, and the device had a sensitivity of 92.9% and specificity of 89.6% for the diagnosis of BRD when the referent diagnostic test in the study was standard thoracic auscultation (Mang et al., 2015). Further research on the diagnostic accuracy and utility of the computer-assisted stethoscope in preweaned, weaned dairy calves and adults is necessary.

3.2 Pathogen detection/sampling techniques

Bacterial isolation from the airway can be useful to define bacterial presence and describe antimicrobial susceptibility profiles and microbial communities. Airway sampling can be used on individual clinical cases, as part of herd monitoring, and for research investigations. Several different techniques for sampling the airway of cattle are possible. Each method has advantages and disadvantages that make optimal method selection dependent on the specific goals of testing. Table 1 summarizes techniques used for obtaining bacterial samples from the airways of cattle. Collection of airway samples from live animals may be more physically challenging than post-mortem collection, depending on the animal size and restraint methods available. However, limited collection of data from dead animals may skew results toward treatment failures and chronic cases. Collection of upper airway samples as a proxy for lower airway infection is relatively technically simple compared to sampling the lower airway, but these methods inherently sample different microbial populations (Nicola et al., 2017; Timsit et al., 2018). However, the pathophysiology of the BRD complex involves changes in bacterial flora of the upper respiratory tract advancing to lower airway infection, so samples from the upper respiratory tract can still be useful in the study of BRD (Doyle et al., 2017b). For all sites, results should be interpreted in light of treatment history and chronicity of disease; these factors may influence the bacteria isolated at the time of sampling and may not necessarily reflect the bacteria present at disease onset. Additionally, sample handling and culture/sensitivity methods employed on respiratory samples should be consistent with Clinical Laboratory Standards Institute (CLSI) guidelines.

3.2.1 Methods review

A nasal swab (NS) is typically performed by simply swabbing the rostral internal surface of the nares with a sterile culturette swab. This type of sampling collects bacteria or other microbes from the rostral nasal cavity and is very simple to

Table 1 Summary of airway bacterial sampling methods in cattle

Sample type	Regional population sampled	Technical notes
Nasal swab (NS)	• Upper airway-rostral nasal cavity	• Technically simple to perform. Requires only temporary restraint. • Maximum contamination of rostral nasal microbiota
Deep nasopharyngeal swab (DNPS)	• Upper airway, nasopharyngeal cavity	• Technically relatively simple to perform, less contamination from rostral nasal cavity than NS, but some contamination of rostral nasal microbiota likely still present
Transtracheal wash (TTW)	• Combination of lower airway and pool of large conducting airways	• Avoids contamination of nasal microbiota • Requires technical skill and thorough restraint
Bronchoalveolar lavage (BAL)	• Combination of lower airway and focal large conducting airways	• Some contamination of nasal microbiota possible (depends on a specific technique) • Requires technical skill and thorough restraint
Thoracocentesis	• Pleural space	• No contamination from internal airways • Requires the presence of pleural effusion; bacterial population in effusion may or may not represent bacteria present in lungs • Requires technical skill and appropriate restraint
Lung tissue culture	• Lower airway, terminal small conducting airways and gas exchange surfaces	• Avoids contamination of nasopharyngeal microbiota • Usually only performed post-mortem • Requires uncontaminated or seared lung specimens

perform. The NS, however, presents a challenge if the goal is the selective isolation of classic respiratory bacteria (*M. haemolytica*, *P. multocida*, *H. somni*); this method contains more contaminants from the rostral nasal cavity, and selective isolation of the desired bacteria may be less successful (Doyle et al., 2017a).

A deep nasopharyngeal swab (DNPS), also known as a guarded nasopharyngeal swab, is performed by using a longer, guarded sterile swab that is passed deeper into the nasal cavity, to the level of the nasopharynx or into the pharyngeal recess, where the guard is removed and the swab is gently rotated to allow sampling of the caudal nasopharynx; the swab is then retracted back into the guard and removed from the nasal cavity. This technique samples upper respiratory microbes while minimizing contamination from the rostral nasal cavity.

The transtracheal wash (TTW) is performed by passing a sterile cannula percutaneously and aseptically into the trachea, followed by passing sterile flexible tubing through the cannula and down the trachea at the ventral most portion before the mainstem bronchi, depositing sterile saline and then immediately aspirating that sterile saline. The resulting fluid contains a wash of tracheal contents representing a pooled lower airway sample of microbes and any exfoliated airway cells. There are variations on this technique used for post-mortem tracheal wash sampling (Godinho et al., 2007). Another variation on the TTW, a transtracheal swab, has been reported (DeRosa et al., 2000).

A bronchoalveolar lavage (BAL) is performed by passing a sterile flexible tube from the nares (or oral cavity) into the trachea to a terminal bronchus, where the tube is gently wedged in a terminal bronchus. Sterile saline is deposited and immediately aspirated through the tube to create a fluid wash from the terminal airway.

A variation on TTW and BAL is the endoscopic guided airway washing sample; this type of procedure is typically performed in a hospital setting in cases when airway endoscopy is indicated. A cannula is typically passed through the endoscope channel, or the endoscope channel is used directly to deposit and aspirate saline into the desired airway location to create a wash of cells and microbes from the lower airway for analysis similar to a TTW or BAL, depending on where the scope/cannula are placed. Care must be taken in the interpretation of samples obtained through the endoscope lumen because contamination from the endoscope is possible.

Thoracentesis is used to sample fluid that has accumulated in the pleural space and is performed by aseptically obtaining a percutaneous sample of pleural effusion. This is usually accomplished using a needle, IV catheter or sterile teat cannula after a small cut down of the skin and intercostal fascia. It is imperative that negative pressure is maintained during thoracentesis to avoid pneumothorax.

Lung tissue can be directly sampled during necropsy; if samples are intended to be used for pathogen identification, the sample should be obtained sterilely, or the tissue section should be seared prior to sampling the internal surface for microbial isolation.

These techniques are often grouped into discussions on upper versus lower airway sampling; NS and DNPS are common methods for upper airway sampling while TTW, BAL (and their variations) and lung tissue culture are typically considered methods for lower airway sampling. Both TTW and BAL techniques involve passing through portions of the upper respiratory tract and inherently contain microbes from the large conducting airways of the trachea or bronchi in addition to what is obtained from lavage of the lower conducting and terminal gas exchange airways. Thus, when discussing lower respiratory tract microbiota sampling ante-mortem, we are technically discussing the sum of microbiota obtained from sampling the terminal conducting airways, transitional airways and gas exchange surfaces. This microbial population is distinct from the population found in the nosopharyngeal cavity, which is typically referred to as the upper respiratory microbiota (Nicola et al., 2017; Timsit et al., 2018). Of these methods, the four most common ante-mortem airway sampling methods in cattle are likely NS, DNPS, TTW and BAL. These methods are typically being used for bacterial culture, identification, and antimicrobial susceptibility testing. More recent studies use these methods to acquire samples for metagenomic analysis. Comparisons between methods for bacterial isolation and identification, including NS, DNPS, TTW and BAL, have been described (Allen et al., 1991; DeRosa et al., 2000; Godinho et al., 2007; Doyle et al., 2017a; Van Driessche et al., 2017). There are conflicting reports comparing agreement between upper and lower airway sampling techniques for bacterial identification on the calf level; on the group level, agreement between methods is generally good (Allen et al., 1991; DeRosa et al., 2000; Godinho et al., 2007). The antimicrobial susceptibility profiles between upper and lower airway samples have been compared, and the two sampling sites reportedly demonstrate the same susceptibility profiles (96%) the vast majority (96%) of the time (DeRosa et al., 2000).

3.3 Bacterial identification

Bacterial identification of BRD pathogens can be obtained by a variety of methods. The technology best suited to identify respiratory bacteria depends on the goals of testing, sample type available and economics. Traditionally, respiratory bacteria have been isolated via culture and identified by physical and biochemical properties. An advantage of traditional culture methods for the identification of common respiratory bacteria is that this technique is commonly readily available at animal disease diagnostic laboratories. The turnaround time for results is usually on the order of days to allow for bacterial growth and identification. This turnaround time is notably longer for *Mycoplasma* species due to their slow-growing biology. More recently, the MALDI-TOF technology has been used for the identification of cultured BRD pathogens. This technology uses mass spectrometry to identify bacteria from cultured samples based on protein structures. This technology is faster than traditional physical and biochemical identification methods and reflects categorization based on the proteome rather than physical or chemical properties. PCR offers another way to identify specific respiratory pathogens. This method uses nucleic acid primers to match for known bacterial pathogens of interest in the sample. Metagenomics is a newer method used to study populations of microbes in the respiratory tract. This technology is used to identify populations of microbes present in a sample rather than selective isolation or specific testing for select organisms. Metagenomics technology works by matching nucleic acid sequences in a given sample to a reference database to identify the microbes present in the sample.

3.4 Determination of antibiotic susceptibility

The use of antibiotic susceptibility testing is one tool we can use to monitor resistance and make antimicrobial stewardship decisions. There are several ways bacterial susceptibility or resistance can be defined. The ultimate measure of interest is often the case outcome, and in vitro methods of susceptibility testing are intended to estimate a prediction of the likelihood of successful bacterial killing in the living animal. There are several methods that can be used, and most commonly include microbroth dilution and disk diffusion methods; microbroth dilution is considered the current gold standard (CLSI, 2018). These tests evaluate bacterial growth inhibition with different concentrations of antimicrobial to identify a minimum inhibitory concentration (MIC). The MIC is compared to CLSI (2018) breakpoints, and the bacterial sample is classified as susceptible, intermediate or resistant. These breakpoints are based on the knowledge of drug pharmacokinetics and disease tissue distribution and are used to predict how likely a given bacterial sample is to respond to treatment with a given drug in the species of interest. The CLSI breakpoints for BRD bacteria and the antibiotics commonly used to treat BRD are available and updated occasionally (CLSI, 2018).

Another way that AMR has been monitored is through molecular techniques to identify known or putative resistance genes. Molecular methods used for the identification of resistance genes in BRD bacteria include PCR, whole genome sequencing (WGS) and metagenomic analysis; some of these methods can also employ quantitative analyses to quantify the amount of resistance gene presence. This information can be useful in tracking and studying AMR and as part of antimicrobial stewardship plans. It should be recognized, however, that AMR phenotypes and resistance genotypes are not synonymous. The presence of a given resistance gene does not necessarily imply that the gene is being expressed and that the bacteria will display resistance to a given antibiotic. Both genotypic and phenotypic tests for antibiotic resistance can be useful for understanding antibiotic resistance; however, it is necessary to recognize the inherent differences in these diagnostic methods when interpreting the results of either type of test.

Case outcomes have also been used as a gauge for antibiotic resistance, although the nature of living systems does not have a perfect agreement with in vitro susceptibility testing (Watts and Sweeney, 2010; Coetzee et al., 2019). A case outcome represents the ultimate measure of antibiotic treatment failure or success; however, many factors influence treatment failure besides specific antibiotic resistance (Watts and Sweeney, 2010). Treatment records that include case outcomes are a valuable tool for on-farm monitoring of AMR patterns.

3.4.1 Importance of antimicrobial susceptibility monitoring

Bacterial causes of disease are commonly treated with antimicrobial drugs, a necessary tool when preventative methods have failed, to support the welfare of our food animals. However, with the use of antibacterial drugs comes the risk of AMR; this resistance affects both cattle and human populations through the selection of resistance genes in the exposed bacterial population. Antibiotic resistance in BRD pathogens has been documented for decades (Chirino-Trejo and Prescott, 1983; Speer et al., 1992) with a concerning trend of increasing resistance over time (CDC, 2019). Bacterial resistance to antibiotics threatens animal, environmental and human health and represents a serious One Health concern. Antimicrobial stewardship, which includes judicious use of antibiotics to reduce the selective pressure for AMR and is necessary to ensure the continued availability of effective antibiotics against infections in dairy cattle, guarantee the health of our herds, ensure a continuous supply of wholesome milk and dairy beef and protect the public health. Surveillance efforts for AMR should involve the bacterial resistance phenotype as well as the genetic determinants of the bacterial resistance. Resistance gene identification can encompass single bacterial strain testing or bacterial population surveillance, including all resistance genes in a population, often referred to as the bacterial resistome. Understanding the complex pathways between on-farm drug use and AMR in dairy cattle populations is key to antibiotic stewardship and judicious use (Mathew et al., 2007). Surveillance should capture the on-farm management practices and bottlenecks to antimicrobial stewardship using both industry-wide and individual farm-specific surveys that investigate the role of staff in antimicrobial drug use, including sources of information, decision processes for acquisition and storage, treatment protocol design and implementation, training and updating of standard protocols on farm and monitoring of treated animals for withdrawal.

There is no consensus describing exactly how to monitor a livestock facility for AMR of BRD pathogens. In the authors' opinions, it seems reasonable to use a multimodal approach. For routine surveillance, culture and sensitivity data can be collected from animals that die or are euthanized due to BRD as well as from acutely affected live animals. Samples can be obtained in the face of a new BRD problem or in groups with ongoing BRD problems for which treatment protocols may need to be updated. Development of objective, goal oriented, clinically useful, AMR monitoring programs are needed as part of a holistic approach to managing BRD.

4 Risk assessment tools

A multiple causation model can be built mirroring the three components of the host, environment and pathogen triangle; such a model incorporates both host and environmental factors as components in the causation of disease in addition to the pathogen disease model traditionally described by Koch's postulates. Failing to incorporate opportunities to modify host and environmental aspects of the disease triangle would limit valuable prevention efforts of bacterial diseases in livestock. Factors discussed earlier, such as environmental factors including heat stress and presence of dust or host factors such as failure of transfer of passive immunity, lack of immunizations or improper nutrition, can all increase the risk of respiratory disease in calves. Hence, the most direct way to break the triad of respiratory disease is to modify the management of calves to mitigate these risk factors. The multifactorial nature of respiratory disease in calves lends itself particularly well for a risk assessment approach for its control and prevention (Gorden and Plummer, 2010). Management on a dairy can be used to limit or modify calf exposure to pathogens and host and environmental risk factors. Developing such a control and prevention plan should start by classifying the risk factors into logical intervention themes, such as colostrum management, housing and vaccination, among others, that contribute to the animals' risk of disease. Hence, a control and prevention plan should be based on a comprehensive assessment of all known risk factors in the herd to allow the investigator to develop a customized herd-specific plan. The resulting herd-specific plan would offer the producer a menu of interventions for control and prevention. Hence, a comprehensive tool that is simple yet functional, based on the results of research on the risk factors associated with the disease of interest, is required so users may elect to implement as many changes in their management as is feasible to reduce the risk of disease. The control and prevention tool should also be rapid and of low cost, allowing the user to benchmark the disease occurrence in their herd repeatedly over time. Benchmarking the disease occurrence before and after the interventions is an implemented act as a demonstration of the impact of control and preventative measures, which may be a positive reinforcement for the producer to continue the control efforts and add more interventions.

Risk assessments are one example of such functional and comprehensive tools to prevent and control diseases and health outcomes of complex causality. Examples of common risk assessment tools in human medicine include the Falls Risk Assessment Tool (FRAT) for fall accidents in seniors (Nandy et al., 2004), the Fracture Risk Assessment Tool (FRAX) (Kanis et al., 2008) and the coronary risk assessment (Pearson, 2002). In veterinary medicine the Johne's disease risk assessment has been used in dairy and beef cattle to control *Mycobacterium avium* subsp. *paratuberculosis*, a disease which, after an elongated incubation period, causes a chronic debilitating enteropathy (Aly and Thurmond, 2005). A risk assessment approach for the prevention of respiratory disease in preweaned calves may be even more successful because producers may observe the benefits of the prescribed interventions relatively soon compared to diseases with longer incubation periods like Johne's disease (USDA, 2011). The following section describes the portions of the BRD risk assessment tool for the prevention of pneumonia in preweaned dairy calves.

4.1 BRD Risk Assessment Tool

Users of the BRD Risk Assessment Tool include specialists and dairy industry personnel, including producers and herd managers of dairies and calf ranches, herd veterinarians, consultants, extension specialists, nutritionists, pharmaceutical companies, government entities (federal and state), researchers and allied industry specialists concerned with dairy cattle health, welfare and productivity. Users can complete the risk assessment tool to benchmark a preweaned calf herd's risk of respiratory disease, identify its risk factors, estimate the herd prevalence and devise a herd-specific management plan to control and prevent respiratory disease. The complete BRD Risk Assessment Tool is available online (https://escholarship.org/uc/item/1jb2f7rm) and is made of three portions described in the next few sections (Aly et al., 2020; Maier et al., 2020).

4.1.1 Risk assessment questionnaire

The risk assessment questionnaire contains questions that collect data on risk factors on a dairy along with a specific risk score for each question's response option. The greater the risk score associated with a practice, the greater the occurrence of respiratory disease. The questions were based on previous research used to gather risk factor information from dairies from three studies. The first study was a mail survey mailed to all grade A milk-producing dairies (Love et al., 2016b). The second study was a cross-sectional study of 100 dairies (BRD 100) distributed along California's three diverse milk sheds (Northern CA, Northern San Joaquin Valley, and the third milk shed composed of both Southern San Joaquin Valley and Southern CA; Love et al., 2016b). The BRD 100 linked management risk factors for over 4253 calves selected randomly from approximately 25 000 calves on these 100 dairies (Maier et al., 2019a). The third study was a longitudinal prospective cohort study of over 11,300 calves followed from birth to weaning on five dairies representing different dairy management systems and allowing for the estimation of the risk of respiratory disease in calves (Dubrovsky et al., 2019a).

The risk assessment questionnaire is made up of the following six sections investigating adherence to best management practices for the prevention of respiratory disease in calves. The sections are basic herd risk profile, maternity pen, colostrum management, calf milk feeding, vaccinations and calf housing. Risk scores from the six sections vary in their total proportion to the magnitude of their risk factor associations with respiratory disease; however, for ease of use, they were scaled so that the entire risk assessment tool scores added up to 1000 (Aly et al., 2020; Maier et al., 2020). The six sections thematically group respiratory disease risk factors, as presented here.

4.1.1.1 Basic herd risk profile

The basic herd risk profile section of the risk assessment characterizes the baseline risk of respiratory disease in the calf herd. The questions in this section refer to herd size, breed distribution, organic status and season and represented 10.3% of the total risk score.

4.1.1.2 Maternity pen management

Risk factors in the maternity pen management include: whether cows and heifers calve together, the frequency of maternity pen bedding changes and the type of bedding including pasture. Regardless of whether calves were born on pasture or not, they had greater odds of respiratory disease if they remained with the dam more than an hour after birth compared to those removed within an hour from birth (Maier et al., 2020). The association between group calving and calf respiratory disease was modified by the frequency of bedding changes. Specifically, if maternity bedding was changed less than six times a month, the occurrence of respiratory disease in the newborn calves increased compared to if the bedding was changed six or more times a month (Maier et al., 2020). In terms of bedding material, the use of dried manure or gypsum was associated with reduced occurrence of respiratory disease in calves compared to dirt or plant materials such as shavings, straw or almond shells. Maternity pen-related risk factors accounted for 10.5% of the total risk score.

4.1.1.3 Colostrum management

The first question in the colostrum section identifies the sources of colostrum as individual dam, pooled, mix of natural and colostrum replacer or nursed from the dam on pasture. Colostrum from an individual dam or colostrum nursed from the dam on pasture was associated with the least odds of respiratory disease (and thus are assigned the lowest risk score), compared to the remaining sources. The remaining questions enquire if colostrum is supplemented or heat-treated and, if so, what is the duration of storage from harvest to heat treatment, is it tested for bacterial content, is it harvested from uniparous dams, is preservatives added, is it stored in solid containers or bags and finally if it is tested for immunoglobulin content. While some of these practices may not have a direct effect on respiratory disease in calves, such as testing colostrum quality, the producers' interventions based on results of testing colostrum for bacteriological content may reduce the risk of respiratory disease in calves. Such interventions may include proper harvest procedures and storage conditions and storage times for colostrum (Maier et al., 2019a). Another example of a risk factor with an indirect effect on the risk of respiratory disease is the necessity of monitoring immunoglobulins of colostrum supplemented with a replacer; the intervention based on a finding of inadequate quality colostrum is what alters the risk of respiratory disease, not the test itself. Other management practices warranted data exploration such as the duration of colostrum storage prior to heat treatment. Specifically, a greater occurrence of respiratory disease was observed when colostrum was stored for more than 10 h before heat treatment, which may be due to higher bacterial concentration compared to when it is stored for 10 or less hours (Maier et al., 2019a). There were lower odds of respiratory disease in calves when dairies stored their colostrum in bags compared to solid containers. Colostrum stored in solid containers may be subject to uneven freezing and thawing, resulting in more damage to the valuable immunoglobulin proteins compared to colostrum stored in bags (Maier et al., 2019a). Testing colostrum for immunoglobulins was associated with a lower risk of respiratory disease in calves, and while testing itself doesn't protect the calves, it is likely the actions taken in response to the test results may protect calves.

The last two questions in the colostrum management section obtain information on the volume of colostrum fed and whether the producer tests calves for the transfer of passive immunity. Higher risk scores are warranted if calves are fed less than 2.84 L (3 US quarts) in the first 12 h after birth or if calves are not tested for the transfer of passive immunity or if they are tested but no actions are taken in the case of failure of the transfer of passive immunity. The colostrum management section accounted for 14.5% of the total risk score.

4.1.1.4 Milk feeding

Questions in the milk feeding section investigate the source of milk fed to the calves, the order of feeding in terms of the age and health status, if milk fed is tested for bacteria and if it is supplemented with a milk replacer. Zero risk score is assigned if dairies feed saleable milk due to its superior nutritional composition compared to milk replacer or a mix of both. Higher risk scores are assigned to dairies that feed calves in no particular order or feed older or sick calves first. Similarly, higher scores are assigned to dairies that feed unpasteurized milk, do not test for bacteria or feed milk supplemented with replacer. The final two questions of the milk feeding section address the volume of milk fed and the daily feeding frequency. While calves fed greater than 3.79 L (4 US quarts) were assigned zero risk scores; the penalty for feeding less was modified by breed due to the differences in body weight and hence nutrient requirement differences between Holsteins and Jersey calves. Specifically, greater risk scores were assigned when Holstein calves were fed less than 3.79 L compared to if the same volume was fed to Jersey calves. The milk feeding section accounted for 24.4% of the total risk score.

4.1.1.5 Vaccination

Despite the lower percent of the total risk scores assigned to the vaccination section (4.4%), targeted vaccination continues to be an important and essential practice to maintain healthy calves (Windeyer et al., 2012; Dubrovsky et al., 2019a; Maier et al., 2019a; Chamorro and Palomares, 2020). The questions in the vaccination section of the risk assessment address whether intranasal vaccines, injectable vaccines either killed or modified live vaccines are administered to calves and if dams are vaccinated. For the latter, higher risk scores are assigned if dams are not vaccinated against respiratory pathogens; however, the penalty is greater for the lack of vaccination using a modified live vaccine compared to the lack of vaccinating using a killed vaccine. The larger penalty for missing the use of a modified live vaccine compared to a killed vaccine relates to the greater reduction in the risk of respiratory disease associated with the use of the former compared to the latter.

4.1.1.6 Housing

Questions related to calf housing management practices represented the largest section of all six by the sum of scores (35.9%). The questions enquired about dust, shade, calf-to-calf contact, housing enclosure type, material and manure removal method. With regard to dust the questions assigned higher risk scores if surfaces surrounding the calf housing were not gravel or paved, owner reports of dust in the calf raising area and for lack of dust abatement.

For the latter, the use of water trucks is assigned higher scores despite it being a common dust abatement method since frequently the same trucks travel at high speeds generating more dust rather than performing dust abatement (Dubrovsky et al., 2019a). Interestingly, the application of magnesium chloride on surfaces between and around calf hutches was found to prevent dust, a technique commonly used in equestrian sports to reduce dust in arenas, dairies that applied this substance received zero risk scores in this field.

With regard to shade, dairies that housed calves under an additional shade structure including the use of shade cloth were assigned zero risk scores. In contrast, dairies with a calf raising area with no roof, with one to three walls with a roof or completely indoors receive higher risk scores. In addition, dairies where calves are housed in groups are assigned a higher risk score compared to dairies that house calves individually. In addition, if calves 75 days old or more have calf-to-calf contact, a higher risk score is assigned compared to if calves of the same age had no calf-to-calf contact.

With regard to housing enclosure type, dairies that utilized calf hutch walls and roof made of a metal material, plastic or a mix of materials all received higher risk scores compared to dairies that house calves in wooden hutches. Dairies that housed calves on raised floors made of plastic-coated wire are also assigned higher risk scores compared to wood floors. Finally, the use of recycled lagoon water to flush manure under the hutches was assigned higher risk scores compared to those that did not flush or flushed with fresh water.

Once the risk assessment questionnaire is completed, investigators can sum the scores from each section's questions and over all the sections to arrive at the herd's risk for respiratory disease by section and overall, respectively. A dairy's risk score can then be benchmarked across other dairies' risk for respiratory disease by section (basic herd risk profile, maternity pen, colostrum management, milk feeding, vaccination and housing) into very low, low, moderate, high or very high risk for respiratory disease based on the percentile distribution of the risk scores (Fig. 4).

4.1.2 The California BRD scoring system

The California (CA) BRD scoring system is used to estimate the prevalence of respiratory disease as a reference point for the specific herd. The CA BRD scoring system has the advantage of being a quick and easy calf-side system that allows for the rapid assessment of calves with minimal handling (Love et al., 2014). The system's accuracy in terms of sensitivity and specificity, like other scoring systems and laboratory-based diagnostic systems, is not perfect (Love et al., 2016a). However, the CA BRD scoring system allows the investigator to compare the prevalence of BRD before and after implementing changes in management to control BRD. Prevalence estimation can be completed using

Risk score section	Bovine Respiratory Disease risk score[1] percentiles[2]				
	10th percentile	25th percentile	50th percentile	75th percentile	90th percentile
Basic Herd risk profile[3]	43	47	63	72	81
Maternity pen management[4]	24	46	65	82	89
Colostrum management[5]	41	58	71	82	96
Milk feeding[6]	52	79	108	142	155
Vaccinations[7]	0	3	7	27	31
Calf housing[8]	106	130	170	199	217
Total score	357	414	467	550	655

[1] Based on the BRD 100 (http://dx.doi.org/10.3168/jds.2018-14773) and BRD 10K studies (https://doi.org/10.3168/jds.2018-14774)
[2] Percentiles based on risk assessment scores for the BRD 100 study herds.
[3] Herd size, organic status, region, season, herd breed composition
[4] Bedding changes maternity pen separate for cows and heifers, bedding type, pasture calving, cows and heifers calve together, % calves removed from dam within an hour
[5] Source, supplementation with colostrum replacer, heat treatment, storage before heat treatment, testing for bacteria, storage container type, volume fed, testing for failure of transfer of passive immunity
[6] Feeding order by age, feeding order by health, source, pasteurization, volume fed, feeding frequency
[7] Intranasal for calves, injectable (killed or modified live) for calves, modified live for cows, killed for cows
[8] Type of surface next to hutches, dust perceived a problem, group housing, extra shade structures, calf-to-calf contact in calves >75 days, hutch wall material, hutch floor material, flush system for manure

Figure 4 Heat map benchmarking bovine respiratory disease (BRD) risk scores. (Image reprinted by permission from Dr. Sharif S. Aly).

an application available for smartphones and tablets called the UC Davis BRD Score. The application is available in English, Spanish and Arabic languages with plans for translating it to more languages. The application can be downloaded from both iOS and Android respective application venues after a keyword search for UC Davis and BRD (Fig. 5).

The CA BRD score application can aid producers in scoring a calf herd. The application requires the calf herd size and then uses a sample size formula to calculate the number of calves required to estimate the prevalence of respiratory disease in the herd, with a bound on the error of estimation that is also user-specified. A user scores the calf in the hutch identified by the app, which also provides the number of hutches to count off from the first hutch on the farm. A similar adaptation can be utilized for calves housed in pens. A user proceeds to the next randomly selected hutch based on the app's output of how many hutches to skip. Between scoring calves, the application enquires whether the calf to score next is present in the hutch since hutches may be vacant as when a calf dies or is removed for other reasons. If the next hutch is empty, the application will instruct the user to proceed to the next hutch (Fig. 6). The application will internally adjust the vector of hutches to score such that the sample size is fulfilled close to but before the last hutch (N, the total number of calves in the herd).

A smart function in the application is that unchecked clinical signs are assumed absent, so users may select only the observed clinical signs by tapping on the example pictures. The app also features a searchable database for calves scored if necessary, using information entered by the user for the scored calves, such as date of scoring, herd, calf identification, breed, sex and age (Fig. 7).

Figure 5 Smart phone application to estimate the prevalence of bovine respiratory disease (BRD) in preweaned dairy calves. (Image reprinted by permission from Dr. Sharif S. Aly).

Once the prevalence estimation is complete, users can email a prevalence report to the herd producer/herd veterinarian for integration into their risk management assessment (Fig. 8).

4.1.3 BRD control and prevention herd management plan

It is based on the completed risk assessment questionnaire; however, the questions are reworded into interventions. The management plan consists of the same sections of the risk assessment questionnaire, with the exception of the basic herd profile section because herd profile factors are mostly static and difficult to change. For example, the first section of the herd management plan in the maternity pen section would be 'Calve cows and heifers separately' if the risk assessment questionnaire had identified that cows and heifers calved together and a risk score was assigned. Once all the interventions are reviewed in light of the risk assessment questionnaire selections, producers

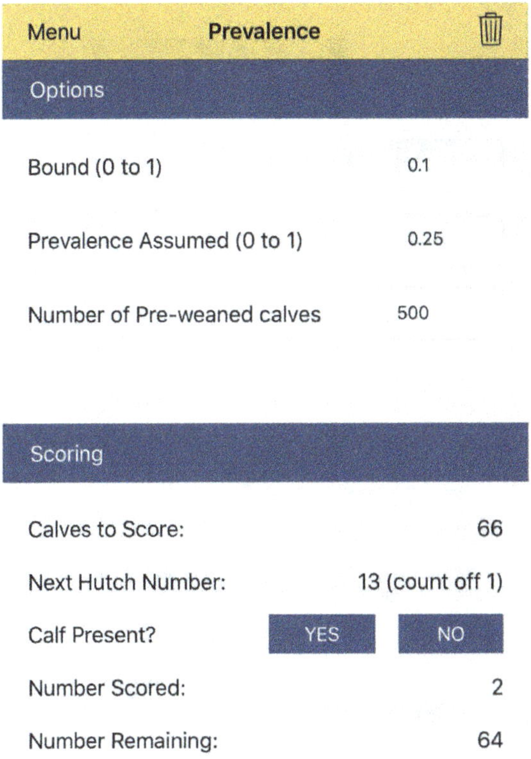

Figure 6 Sample size inputs in the California bovine respiratory disease score application to estimate the prevalence of BRD in preweaned dairy calves. (Image reprinted by permission from Dr. Sharif S. Aly).

can then elect to apply any or all of these interventions finalizing their herd-specific management plan. Producers are advised to identify feasible changes in terms of their available staff effort and financial resources. The decision regarding which feasible interventions produce the greatest impact on the herd's respiratory disease risk should be based on the resulting reduction in the herd's risk score. For example, the investigator may preferentially implement interventions related to risk factors from one component over another because of a greater reduction in the overall risk score. Producers should also consider the distribution of respiratory cases by age, breed, calf sex and clinical signs, which may be easily identified using the CA BRD score application. Although the clinical signs for BRD are not specific, an assessment of frequent scores may provide clues to specific underlying causes, such as tilted heads or drooped ears, which are commonly associated with *M. bovis* infections. Similarly, the

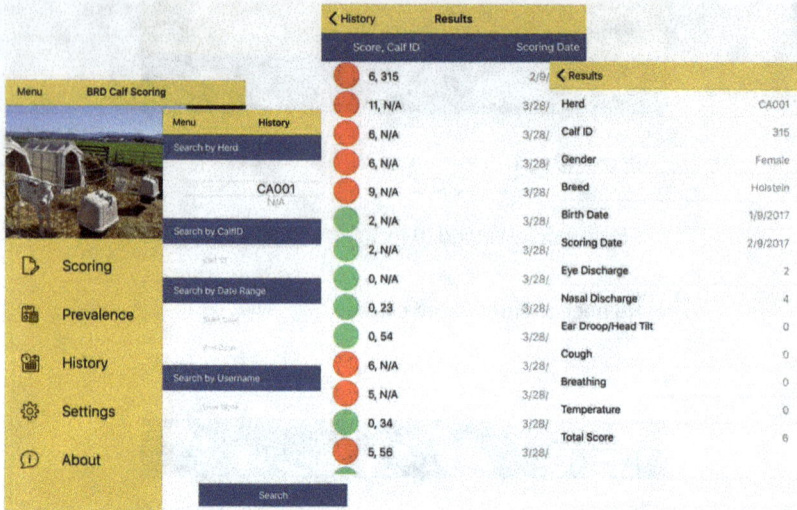

Figure 7 Database search features in the California bovine respiratory disease score application to estimate the prevalence of BRD in preweaned dairy calves. (Image reprinted by permission from Dr. Sharif S. Aly).

age distribution of cases may help narrow down underlying management problems. For example, if many respiratory cases are seen in neonatal calves, one could investigate aspiration pneumonia commonly seen in the first few days of life if calves are tube fed by untrained staff. An increase in cases in one breed more than another, in mixed breed herds, may be related to breed-associated management differences such as the use of a specific hutch type or volume of milk fed for one breed but not another. Differences in respiratory disease between males and females may be confounded due to practices such as feeding female calves more colostrum than male calves.

Once the feasible changes are identified, the implementation plan should detail the staff in charge of each intervention and an implementation plan following a timetable with deadlines. Such details identify responsibilities and allow the calf caretakers, producer and investigator to work as a team to identify bottlenecks and hurdles, particularly if the intervention changes agreed upon do not take place. Producers can then re-estimate the prevalence of BRD after implementing the recommended management changes and elapse of a suitable period to allow for a new cohort of calves to have experienced these changes.

5 Future trends in research

It is imperative that we consider all factors of the disease triangle when forming disease prevention plans. Table 2 illustrates some examples of common BRD

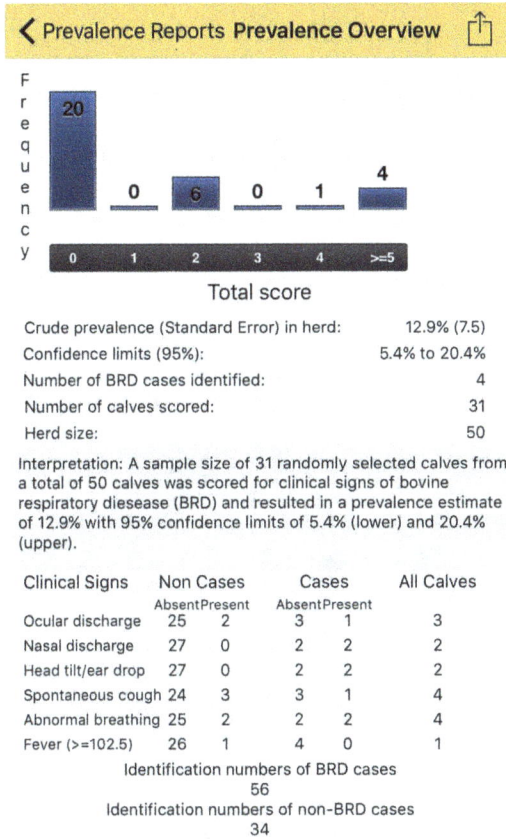

Figure 8 Prevalence report feature in the California bovine respiratory disease score application to estimate the prevalence of BRD in preweaned dairy calves. (Image reprinted by permission from Dr. Sharif S. Aly).

modifying factors with the related host, environment and pathogen factors and basic intervention recommendations. As agricultural practices evolve so do the disease syndromes of livestock. The assumptions used to create the recommendations within this chapter will likely change over time as modern agriculture continues to evolve. Animal health interventions must remain dynamic and responsive to the state of animal agriculture; interventions must be constantly measured, monitored for success and re-evaluated. As animal health professionals, we not only react to animal disease events but must be proactive to influence change in animal management standards.

Table 2 Summary of common BRD modifiers and the related host, environment and pathogen factors with feasible intervention recommendations

Modifier	Host, environment, pathogen factors			Recommended action	References
	Host	Environment	Pathogen		
Physiologic and psychological stress	• Stimulation of the HPA axis and SNS • Blunted immune responses	• Shipping, processing environment • Other human handling • Changes in animal groups	• Shifts in respiratory microbial commensal population • Possible increases in pathogen virulence	• Minimize physical stressors (processing, shipping, etc.) • Minimize psychological stressors such as social group changes and stressful handling • Acclimate calves to human handling	Carroll and Forsberg (2007), Mansfield et al. (2008), Ackermann et al. (2010), Lyte (2014b)
Heat stress	• Increased energy use • Decreased feed intake • Inflammation • Impaired respiratory immunity	• High ambient temperatures or temperature humidity index • Fluctuations in temperature		• Pay attention to both macro and micro environment of animal housing (ex: ambient and temperature within calf hutches) • Employ heat abatement techniques such as increasing shade	Jones (1987), Lammers et al. (1996), Mitlöhner et al. (2002); Burgos et al. (2010), Bertagnon et al. (2011), Pearce et al. (2013a,b), Roland et al. (2016), Kim et al. (2018), Louie et al. (2018), Koch et al. (2019b)

Housing

• Maternity bedding/hygiene	Increased odds of BRD suspect due to high pathogen load in neonatal period	• Dirty maternity bedding • Lagoon water use to flush under hutches	High pathogen load	• Change maternity bedding >5 times a month • Avoid dirt or plant material bedding	Dubrovsky et al. (2019a, b), Karle et al. (2019), Maier et al. (2020), Karle et al. (2019), Maier et al. (2020)
• Hutch construction	Increased odds of BRD, suspect due to heat stress	Metal hutch components, suspect increase heat conduction		• Avoid metal components in hutches in warm climates	Dubrovsky et al. (2019b), Karle et al. (2019), Maier et al. (2020)
• Dust	Increased risk of BRD likely due to the effect of particulate matter and inhaled endotoxin on respiratory anatomy	Dusty environment, traffic on dusty roads near animal housing	Potential for exposure to bacteria	• Employ dust abatement; ensure truck movement does not create dust near animal housing	Purdy et al. (2002), Dubrovsky et al. (2019a), Karle et al. (2019)
• Commingling	Increased contact between calves	Shared microbiome	Exposure to respiratory pathogens	Individual housing, reduce the density in group pens, vaccination against BRD pathogens before commingling	Abdelfattah et al. (2013), Karle et al. (2019), Maier et al. (2019a, 2020)
Nutrition	• Feeding below nutritional requirements is associated with immune dysfunction and an increased risk of BRD • Feeding whole milk (vs milk replacer) is associated with lower risk of BRD	• Heat and cold stress increases nutrient requirements	• Increase in nutrient requirements as the immune system is challenged by infectious agents	• Feed calves optimum diet in terms of composition and nutrient requirements	Griebel et al. (1987), Godden et al. (2005), Drackley (2008), Dubrovsky et al. (2019a,b), Karle et al. (2019), Maier et al. (2019b, 2020)

(Continued)

Table 2 (*Continued*)

	Host, environment, pathogen factors				
Modifier	Host	Environment	Pathogen	Recommended action	References
Vaccination	Build host adaptive immunity to specific pathogens	Possible decrease in environmental/ group pathogen load due to herd immunity	Pathogenicity altered or interrupted by host adaptive immunity	Vaccinate healthy animals for specific pathogens based on farm risk. Adjust vaccine timing to precede risk period	Larson and Step (2012), Chamorro and Palomares (2020), Richeson and Falkner (2020)
Antibiotic resistance	Treatment failures	Contamination of the environment with resistance genes	Decreased efficacy of bacterial killing with antibiotic treatment	Implement an antibiotic use stewardship plan that includes: clear staff training materials for appropriate drug use – review and update regularly using evidence-based medicine for drug selection; clear record-keeping for monitoring drug use and case outcomes; implement antibiotic susceptibility testing and monitoring in representative samples; employ disease prevention methods and limit antibiotic use; consider non-antibiotic treatments where possible	AVMA, Mathew et al. (2007); AABP (2017); Mathew et al. (2007)

6 Where to look for further information

The disease prevention and control concepts outlined in this chapter are presented using BRD as an example and culminate in the form of a herd risk assessment tool. The BRD tool is available online at:

- https://escholarship.org/uc/item/1jb2f7rm.

The approach presented here is relevant to, and has been implemented for, prevention and control of other diseases. For example, the control and prevention efforts for paratuberculosis in cattle, which make up the Johne's disease risk assessment tool. The Johne's disease risk assessment tool precedes the BRD risk assessment and is available in the form of separate tools for both beef and dairy cattle production systems. The risk assessment for Johne's disease is available for beef production systems here:

- https://johnes.org/wp-content/uploads/2018/11/Handbook-for-Vets-and -Beef-Producers-4th-ed-2011.pdf.

Similarly, the risk assessment for Johne's disease is available for dairy cattle here:

- https://johnes.org/wp-content/uploads/2018/11/Handbook-for-Vets-and -Dairy-Producers-4th-ed-2011.pdf.

Even with a successful prevention and control program, bacterial infections still occur. When treating and managing bacterial infections in cattle, it is important to use up-to-date veterinary practices and follow local and regional antimicrobial drug use regulations. A helpful resource for residue avoidance for antimicrobial drugs and other medications in food animals is the Food Animal Drug Residue Avoidance Databank (FARAD): http://www.farad.org/.

Antimicrobial stewardship and residue avoidance practices are important to prevent antimicrobial resistance. Antimicrobial resistance is of global concern and appropriate antimicrobial drug use is essential to preserve the effectiveness of these valuable chemicals for the health and welfare of our food animals and public health. Information on the threat of antimicrobial resistance can be found in the Centers for Disease Control and Prevention, and the World Health Organization reports. As of this chapter's publication date the aforementioned reports can be accessed at:

- https://www.cdc.gov/drugresistance/pdf/threats-report/2019-ar-threats- report-508.pdf.
- https://www.who.int/publications/i/item/global-action-plan-on-antimicr obial-resistance.

7 References

AABP 2017. Key elements for implementing antimicrobial stewardship plans in bovine veterinary practices working with beef and dairy operations. Available online: https://www.aabp.org/resources/AABP_Guidelines/AntimicrobialStewardship-7.27.17.pdf.

Abdelfattah, E. M., Schutz, M. M., Lay, D. C., Marchant-Forde, J. N. and Eicher, S. D. 2013. Effect of group size on behavior, health, production, and welfare of veal calves. *J. Anim. Sci.* 91(11):5455–5465. doi:10.2527/jas.2013-6308.

Ackermann, M. R., Derscheid, R. and Roth, J. A. 2010. Innate immunology of bovine respiratory disease. *Vet. Clin. North Am. Food Anim. Pract.* 26(2):215–228. doi:10.1016/j.cvfa.2010.03.001.

Adams, E. A. and Buczinski, S. 2016. Short communication: ultrasonographic assessment of lung consolidation postweaning and survival to the first lactation in dairy heifers. *J. Dairy Science* 99(2):1465–1470. doi:10.3168/jds.2015-10260.

Adlam, C. and Rutter, J. 1989. *Pasteurella and Pasteurellosis.*

Allen, J. W., Viel, L., Bateman, K. G., Rosendal, S., Shewen, P. E. and Physick-Sheard, P. 1991. The microbial flora of the respiratory tract in feedlot calves: associations between nasopharyngeal and bronchoalveolar lavage cultures. *Can. J. Vet. Res.* 55(4):341–346.

Aly, S. S., Karle, B. M., Williams, D. R., Maier, G. U. and Dubrovsky, S. 2020. Components of a risk assessment tool for prevention and control of bovine respiratory disease in preweaned dairy calves. *Anim Health Res Rev*:1–7. https://doi.org/10.1017/S1466252320000201.

Aly, S. S. and Thurmond, M. C. 2005. Evaluation of *Mycobacterium avium* subsp *paratuberculosis* infection of dairy cows attributable to infection status of the dam. *J. Am. Vet. Med. Assoc.* 227(3):450–454.

AVMA 2020. Antimicrobial Stewardship Definition and Core Principles | American Veterinary Medical Association. Accessed August 10, 2020. https://www.avma.org/resources-tools/avma-policies/antimicrobial-stewardship-definition-and-core-principles.

Baumgard, L. H., Wheelock, J. B., Sanders, S. R., Moore, C. E., Green, H. B., Waldron, M. R. and Rhoads, R. P. 2011. Postabsorptive carbohydrate adaptations to heat stress and monensin supplementation in lactating Holstein cows1. *J. Dairy Sci.* 94(11):5620–5633. doi:10.3168/jds.2011-4462.

Beier, T., Hotzel, H., Lysnyansky, I., Grajetzki, C., Heller, M., Rabeling, B., Yogev, D. and Sachse, K. 1998. Intraspecies polymorphism of vsp genes and expression profiles of variable surface protein antigens (Vsps) in field isolates of *Mycoplasma bovis. Vet. Microbiol.* 63(2–4):189–203. doi:10.1016/S0378-1135(98)00238-7.

Bertagnon, H. G., Esper, G. V. Z., Emanuelli, M. P. and Pel-legrine, L. G. 2011. Nfluência meteorológica no leucograma e na popula-ção citológica do trato respiratório de bezerros. *Pesq. Vet. Bras.* 31(3):244–246.

Blecha, F., Boyles, S. L. and Riley, J. G. 1984. Shipping suppresses lymphocyte blastogenic responses in Angus and Brahman X Angus feeder calves. *J. Anim. Sci.* 59(3):576–583. doi:10.2527/jas1984.593576x.

Boyce, J. D. and Adler, B. 2000. The Capsule is a Virulence Determinant in the Pathogenesis of *Pasteurella multocida* M1404 (B:2). https://doi.org/10.1128/IAI.68.6.3463-3468.2000.

Buchenau, I., Poumarat, F., Le Grand, D., Linkner, H., Rosengarten, R., Hewicker-Trautwein, M. 2010. Expression of *Mycoplasma bovis* variable surface membrane proteins in the respiratory tract of calves after experimental infection with a clonal variant of *Mycoplasma bovis* type strain PG45. *Res. Vet. Sci.* 89:223–229.

Buczinski, S., Forté, G., Francoz, D. and Bélanger, A. M. 2014. Comparison of thoracic auscultation, clinical score, and ultrasonography as indicators of bovine respiratory disease in preweaned dairy calves. *J. Vet. Intern. Med.* 28(1):234–242. doi:10.1111/jvim.12251.

Burgos, S. A., Embertson, N. M., Zhao, Y., Mitloehner, F. M., DePeters, E. J. and Fadel, J. G. 2010. Prediction of ammonia emission from dairy cattle manure based on milk urea nitrogen: relation of milk urea nitrogen to ammonia emissions. *J. Dairy Sci.* 93(6):2377–2386. doi:10.3168/jds.2009-2415.

Calf Health Module–The Dairyland Initiative 2020. Accessed July 31, 2020. https://thedairylandinitiative.vetmed.wisc.edu/home/calf-health-module/.

Callan, R. J. and Garry, F. B. 2002. Biosecurity and bovine respiratory disease. *Vet. Clin. North Am. Food Anim. Pract.* 18(1):57–77. doi:10.1016/S0749-0720(02)00004-X.

Carroll, J. A. and Forsberg, N. E. 2007. Influence of stress and nutrition on cattle immunity. *Vet. Clin. North Am. Food Anim. Pract.* 23(1):105–149. doi:10.1016/j.cvfa.2007.01.003.

Carter, B. H., Friend, T. H., Garey, S. M., Sawyer, J. A., Alexander, M. B. and Tomazewski, M. A. 2014. Efficacy of reflective insulation in reducing heat stress on dairy calves housed in polyethylene calf hutches. *Int. J. Biometeorol.* 58(1):51–59.

Casas, E., Garcia, M. D., Wells, J. E. and Smith, T. P. L. 2011. Association of single nucleotide polymorphisms in the ANKRA2 and CD180 genes with bovine respiratory disease and presence of *Mycobacterium avium* subsp. *paratuberculosis*(1). *Anim. Genet.* 42(6):571–577. doi:10.1111/j.1365-2052.2011.02189.x.

Caswell, J. L. 2014. Failure of respiratory defenses in the pathogenesis of bacterial pneumonia of cattle. *Vet. Pathol.* 51(2):393–409. doi:10.1177/0300985813502821.

CDC 2019. Antibiotic resistance threats in the United States. *Centers Dis. Control. Prev.*:1–150. https://www.cdc.gov/drugresistance/pdf/threats-report/2019-ar-threats-report-508.pdf.

Chamorro, M. F. and Palomares, R. A. 2020. Bovine respiratory disease vaccination against viral pathogens: modified-live versus inactivated antigen vaccines, intranasal versus parenteral, what is the evidence? *Vet. Clin. North Am. Food Anim. Pract.* 36(2):461–472. doi:10.1016/j.cvfa.2020.03.006.

Chigerwe, M., Hagey, J. V. and Aly, S. S. 2015. Determination of neonatal serum immunoglobulin G concentrations associated with mortality during the first 4 months of life in dairy heifer calves. *J. Dairy Res.* 82(4):400–406. doi:10.1017/S0022029915000503.

Chigerwe, M., Tyler, J. W., Schultz, L. G., Middleton, J. R., Steevens, B. J., Spain, J. N. 2008. Effect of colostrum administration by use of oroesophageal intubation on serum IgG concentrations in Holstein bull calves. *Am. J. Vet. Res.* 69(9):1158–1163. doi:10.2460/ajvr.69.9.1158.

Chirino-Trejo, J. M. and Prescott, J. F. 1983. The identification and antimicrobial susceptibility of anaerobic bacteria from pneumonic cattle lungs. *Can. J. Comp. Med.* 47(3):270–275.

CLSI 2018. *Performance Standards for Antimicrobial Disk and Dilution Susceptibility Tests for Bacteria Isolated From Animals; Approved Standard–Fourth Edition.* CLSI document VET01-A4. Wayne, PA: Clinical and Laboratory Standards Institute.

Coetzee, J. F., Magstadt, D. R., Sidhu, P. K., Follett, L., Schuler, A. M., Krull, A. C., Cooper, V. L., Engelken, T. J., Kleinhenz, M. D. and O'Connor, A. M. 2019. Association between antimicrobial drug class for treatment and retreatment of bovine respiratory disease (BRD) and frequency of resistant BRD pathogen isolation from veterinary diagnostic laboratory samples. *PLoS ONE* 14(12):e0219104. doi:10.1371/journal. pone.0219104.

Confer, A. W. and Ayalew, S. 2018. Mannheimia haemolytica in bovine respiratory disease: immunogens, potential immunogens, and vaccines. *Anim. Heal. Res. Rev.* 19(2):79-99. doi:10.1017/S1466252318000142.

Constable, P. D., Hinchcliff, K. W., Done, S. H. and Grünberg, W. 2016. *Veterinary Medicine*. 11th ed. Elsevier.

Corbeil, L. B. 1996. *Histophilus somni* host-parasite relationships. *Anim. Heal. Res. Rev.* 8:151-160. doi:10.1017/S1466252307001417.

Dabo, S. M., Taylor, J. D. and Confer, A. W. 2008. *Pasteurella multocida* and bovine respiratory disease. *Anim. Heal. Res. Rev.* 8:129-150. doi:10.1017/ S1466252307001399.

Davis, C. L. and Drackley, J. K. 1998. *The Development, Nutrition, and Management of the Young Calf*. Ames, IA: Iowa State University Press.

DeRosa, D. C., Mechor, G. D., Staats, J. J., Chengappa, M. M. and Shryock, T. R. 2000. Comparison of *Pasteurella* spp. simultaneously isolated from nasal and transtracheal swabs from cattle with clinical signs of bovine respiratory disease. *J. Clin. Microbiol.* 38(1):327-332.

Downey, E. D., Tait, R. G., Jr, Mayes, M. S., Park, C. A., Ridpath, J. F., Garrick, D. J. and Reecy, J. M. 2013. An evaluation of circulating bovine viral diarrhea virus type 2 maternal antibody level and response to vaccination in Angus calves. *J. Anim. Sci.* 91(9):4440-4450.

Doyle, D., Credille, B., Lehenbauer, T. W., Berghaus, R., Aly, S. S., Champagne, J., Blanchard, P., Crossley, B., Berghaus, L., Cochran, S. and Woolums, A. 2017. Agreement among 4 sampling methods to identify respiratory pathogens in dairy calves with acute bovine respiratory disease. *J. Vet. Intern. Med.* 31(3):954-959. doi:10.1111/ jvim.14683.

Drackley, J. K. 2008. Calf nutrition from birth to breeding. *Vet. Clin. North Am. Food Anim. Pract.* 24(1):55-86. doi:10.1016/j.cvfa.2008.01.001.

Dubrovsky, S. A., Van Eenennaam, A. L., Karle, B. M., Rossitto, P. V., Lehenbauer, T. W. and Aly, S. S. 2019a. Epidemiology of bovine respiratory disease (BRD) in preweaned calves on California dairies: the BRD 10K study. *J. Dairy Sci.* 102(8):7306-7319. doi:10.3168/jds.2018-14774.

Dubrovsky, S. A., Van Eenennaam, A. L., Karle, B. M., Rossitto, P. V., Lehenbauer, T. W. and Aly, S. S. 2019b. Bovine respiratory disease (BRD) cause-specific and overall mortality in preweaned calves on California dairies: the BRD 10K study. *J. Dairy Sci.* 102(8):7320-7328 doi:10.3168/jds.2018-15463.

Dunn, T. R., Ollivett, T. L., Renaud, D. L., Leslie, K. E., LeBlanc, S. J., Duffield, T. F. and Kelton, D. F. 2018. The effect of lung consolidation, as determined by ultrasonography, on first-lactation milk production in Holstein dairy calves. *J. Dairy Sci.* 101(6):5404-5410. doi:10.3168/jds.2017-13870.

Ellis, J., Gow, S., Bolton, M., Burdett, W. and Nordstrom, S. 2014. Inhibition of priming for bovine respiratory syncytial virus-specific protective immune responses following parenteral vaccination of passively immune calves. *Can. Vet. J.* 55(12):1180 -1185.

Ellis, J., West, K., Cortese, V., Konoby, C. and Weigel, D. 2001. Effect of maternal antibodies on induction and persistence of vaccine-induced immune responses against bovine viral diarrhea virus type II in young calves. *J. Am. Vet. Med. Assoc.* 219(3):351-356. doi:10.2460/javma.2001.219.351.

Francoz, D., Buczinski, S., Bélanger, A. M., Forté, G., Labrecque, O., Tremblay, D., Wellemans, V. and Dubuc, J. 2015. Respiratory pathogens in Québec dairy calves and their relationship with clinical status, lung consolidation, and average daily gain. *J. Vet. Intern. Med.* 29(1):381-387. doi:10.1111/jvim.12531.

Gaeta, N. C., Lima, S. F., Teixeira, A. G., Ganda, E. K., Oikonomou, G., Gregory, L. and Bicalho, R. C. 2017. Deciphering upper respiratory tract microbiota complexity in healthy calves and calves that develop respiratory disease using shotgun metagenomics. *J. Dairy Sci.* 100(2):1445-1458. doi:10.3168/jds.2016-11522.

Galyean, M. L., Perino, L. J. and Duff, G. C. 1999. Interaction of Cattle Health/Immunity and Nutrition. *Journal of Animal Science,* 77(5): 1120-1134. https://doi.org/ 10.2527/1999.7751120x

Gardner, I. A., Wong, S. J., Ferraro, G. L., Balasuriya, U. B., Hullinger, P. J., Wilson, W. D., Shi, P. Y. and MacLachlan, N. J. 2007. Incidence and effects of West Nile virus infection in vaccinated and unvaccinated horses in California. *Vet. Res.* 38(1):109-116. doi:10.1051/vetres:2006045.

Gerber, J. D., Marron, A. E. and Kucera, C. J. 1978. Local and systemic cellular and antibody immune responses of cattle to infectious bovine rhinotracheitis virus vaccines administered intranassally or intramuscularly. *Am. J. Vet. Res.* 39(5):753-760.

Gioia, J., Qin, X., Jiang, H., Clinkenbeard, K., Lo, R., Liu, Y., Fox, G. E., Yerrapragada, S., McLeod, M. P., McNeill, T. Z., Hemphill, L., Sodergren, E., Wang, Q., Muzny, D. M., Homsi, F. J., Weinstock, G. M. and Highlander, S. K. 2006. The genome sequence of Mannheimia haemolytica A1: insights into virulence, natural competence, and Pasteurellaceae phylogeny. *J. Bacteriol.* 188(20):7257-7266. doi:10.1128/ JB.00675-06.

Glass, E. J., Baxter, R., Leach, R. J. and Jann, O. C. 2012. Genes controlling vaccine responses and disease resistance to respiratory viral pathogens in cattle. *Vet. Immunol. Immunopathol.* 148(1-2):90-99. doi:10.1016/j.vetimm.2011.05.009.

Godden, S. M., Fetrow, J. P., Feirtag, J. M., Green, L. R. and Wells, S. J. 2005. Economic analysis of feeding pasteurized nonsaleable milk versus conventional milk replacer to dairy calves. *J. Am. Vet. Med. Assoc.* 226(9):1547-1554. doi:10.2460/ javma.2005.226.1547.

Godden, S. M., Smolenski, D. J., Donahue, M., Oakes, J. M., Bey, R., Wells, S., Sreevatsan, S., Stabel, J. and Fetrow, J. 2012. Heat-treated colostrum and reduced morbidity in preweaned dairy calves: results of a randomized trial and examination of mechanisms of effectiveness. *J. Dairy Sci.* 95(7):4029-4040. doi:10.3168/jds.2011 -5275.

Godden, S. M. M., Haines, D. M. M., Konkol, K. and Peterson, J. 2009. Improving passive transfer of immunoglobulins in calves. II: Interaction between feeding method and volume of colostrum fed. *J. Dairy Sci.* 92(4):1758-1764. doi:10.3168/jds.2008-1847.

Godinho, K. S., Sarasola, P., Renoult, E., Tilt, N., Keane, S., Windsor, G. D., Rowan, T. G. and Sunderland, S. J. 2007. Use of deep nasopharyngeal swabs as a predictive diagnostic method for natural respiratory infections in calves. *Vet. Rec.* 160(1):22–25. doi:10.1136/vr.160.1.22.

Gorden, P. J. and Plummer, P. 2010. Control, management, and prevention of bovine respiratory disease in dairy calves and cows. *Vet. Clin. North Am. Food Anim. Pract.* 26(2):243–259. doi:10.1016/j.cvfa.2010.03.004.

Griebel, P. J., Schoonderwoerd, M. and Babiuk, L. A. 1987. Ontogeny of the immune response: effect of protein energy malnutrition in neonatal calves. *Can. J. Vet. Res.* 51(4):428–435.

Griffin, D., Chengappa, M. M. M., Kuszak, J. and McVey, D. S. 2010. Bacterial pathogens of the bovine respiratory disease complex. *Vet. Clin. North Am. Food Anim. Pract.* 26(2):381–394. doi:10.1016/j.cvfa.2010.04.004.

Grooms, D. L., Brock, K. V. and Ward, L. A. 1998. Detection of cytopathic bovine viral diarrhea virus in the ovaries of cattle following immunization with a modified live bovine viral diarrhea virus vaccine. *J. Vet. Diagn. Investig.* 10(2):130–134. doi:10.1177/104063879801000202.

Guo, Y., McMullen, C., Timsit, E., Hallewell, J., Orsel, K., van der Meer, F., Yan, S. and Alexander, T. W. 2020. Genetic relatedness and antimicrobial resistance in respiratory bacteria from beef calves sampled from spring processing to 40 days after feedlot entry. *Vet. Microbiol.* 240:108478. doi:10.1016/j.vetmic.2019.108478.

Hall, J. A., Isaiah, A., Estill, C. T., Pirelli, G. J. and Suchodolski, J. S. 2017. Weaned beef calves fed selenium-biofortified alfalfa hay have an enriched nasal microbiota compared with healthy controls. *PLoS ONE* 12(6):e0179215. doi:10.1371/journal.pone.0179215.

Hammer, C. J., Quigley, J. D., Ribeiro, L. and Tyler, H. D. 2004. Characterization of a colostrum replacer and a colostrum supplement containing IgG concentrate and growth factors. *J. Dairy Sci.* 87(1):106–111. doi:10.3168/jds.S0022-0302(04)73147-1.

Hanthorn, C. J., Dewell, R. D., Cooper, V. L., Frana, T. S., Plummer, P. J., Wang, C. and Dewell, G. A. 2014. Randomized clinical trial to evaluate the pathogenicity of *Bibersteinia trehalosi* in respiratory disease among calves. *BMC Vet. Res.* 10:89. doi:10.1186/1746-6148-10-89.

Hause, B. M., Ducatez, M., Collin, E. A., Ran, Z., Liu, R., Sheng, Z., Armien, A., Kaplan, B., Chakravarty, S., Hoppe, A. D., Webby, R. J., Simonson, R. R. and Li, F. 2013. Isolation of a novel swine influenza virus from Oklahoma in 2011 which is distantly related to human influenza C viruses. *PLoS Pathog.* 9(2):e1003176. doi:10.1371/journal.ppat.1003176.

Hillman, P., Gebremedhin, K. and Warner, R. 2010. Ventilation system to minimize airborne bacteria, Dust, humidity, and ammonia in calf nurseries. *J. Dairy Sci.* 75(5):1305–1312. doi:10.3168/jds.S0022-0302(92)77881-3.

Hoff, J. L., Decker, J. E., Schnabel, R. D., Seabury, C. M., Neibergs, H. L. and Taylor, J. F. 2019. QTL-mapping and genomic prediction for bovine respiratory disease in U.S. Holsteins using sequence imputation and feature selection. *BMC Genomics* 20(1):555. doi:10.1186/s12864-019-5941-5.

Holman, D. B., Timsit, E. and Alexander, T. W. 2015. The nasopharyngeal microbiota of feedlot cattle. *Sci. Rep.* 5:15557. doi:10.1038/srep15557.

Holman, D. B., Timsit, E., Amat, S., Abbott, D. W., Buret, A. G. and Alexander, T. W. 2017. The nasopharyngeal microbiota of beef cattle before and after transport to a feedlot. *BMC Microbiol* 17(1):70. doi:10.1186/s12866-017-0978-6.

Holman, D. B., Timsit, E., Booker, C. W. and Alexander, T. W. 2018. Injectable antimicrobials in commercial feedlot cattle and their effect on the nasopharyngeal microbiota and antimicrobial resistance. *Vet. Microbiol.* 214:140–147. doi:10.1016/j.vetmic.2017.12.015.

Holman, D. B., Yang, W. and Alexander, T. W. 2019. Antibiotic treatment in feedlot cattle: a longitudinal study of the effect of oxytetracycline and tulathromycin on the fecal and nasopharyngeal microbiota. *Microbiome* 7(1):86. doi:10.1186/s40168-019-0696-4.

Howard, C. J. and Taylor, G. 1983. Interaction of mycoplasmas and phagocytes. *Yale J. Biol. Med.* 56(5–6):643–648.

Howard, C. J., Thomas, L. H. and Parsons, K. R. 1987. Comparative pathogenicity of *Mycoplasma bovis* and *Mycoplasma dispar* for the respiratory tract of calves. *Isr. J. Med. Sci.* 23(6):621–624.

Ilg, T. 2017. Investigations on the molecular mode of action of the novel immunostimulator ZelNate: activation of the cGAS-STING pathway in mammalian cells. *Mol. Immunol.* 90:182–189. doi:10.1016/j.molimm.2017.07.013.

Johnston, D., Earley, B., Cormican, P., Murray, G., Kenny, D. A., Waters, S. M., McGee, M., Kelly, A. K. and McCabe, M. S. 2017. Illumina MiSeq 16S amplicon sequence analysis of bovine respiratory disease associated bacteria in lung and mediastinal lymph node tissue. *BMC Vet. Res.* 13(1):118. doi:10.1186/s12917-017-1035-2.

Jones, C. D. 1987. Proliferation of Pasteurella haemolytica in the calf respiratory tract after an abrupt change in climate. *Res. Vet. Sci.* 42(2):179–186.

Jericho, K. W., Lejeune, A. and Tiffin, G. B. 1986. Bovine herpesvirus-1 and *Pasteurella haemolytica* aerobiology in experimentally infected calves. *Am. J. Vet. Res.* 47(2):205–209.

Kainer, R. A. and Will, D. A. 1981. Morphophysiologic bases for the predisposition of the bovine lung to bronchial pneumonia. *Prog. Clin. Biol. Res.* 59B:311–317.

Kanis, J. A., Johnell, O., Oden, A., Johansson, H. and McCloskey, E. 2008. FRAX and the assessment of fracture probability in men and women from the UK. *Osteoporos Int* 19(4):385–397 doi:10.1007/s00198-007-0543-5.

Karle, B. M., Maier, G. U., Love, W. J., Dubrovsky, S. A., Williams, D. R., Anderson, R. J., Van Eenennaam, A. L., Lehenbauer, T. W. and Aly, S. S. 2019. Regional management practices and prevalence of bovine respiratory disease in California's preweaned dairy calves. *J. Dairy Sci.* 102(8):7583–7596. doi:10.3168/jds.2018-14775.

Katsuda, K., Kamiyama, M., Kohmoto, M., Kawashima, K., Tsunemitsu, H. and Eguchi, M. 2008. Serotyping of *Mannheimia haemolytica* isolates from bovine pneumonia: 1987–2006. *Vet. J.* 178(1):146–148. doi:10.1016/j.tvjl.2007.07.019.

Kim, W. S., Lee, J. S., Jeon, S. W., Peng, D. Q., Kim, Y. S., Bae, M. H., Jo, Y. H. and Lee, H. G. 2018. Correlation between blood, physiological and behavioral parameters in beef calves under heat stress. *Asian-Australas. J. Anim. Sci.* 31(6):919–925. doi:10.5713/ajas.17.0545.

Klima, C. L., Holman, D. B., Ralston, B. J., Stanford, K., Zaheer, R., Alexander, T. W. and McAllister, T. A. 2019. Lower respiratory tract microbiome and resistome of bovine respiratory disease mortalities. *Microb. Ecol.* 78(2):446–456. doi:10.1007/s00248-019-01361-3.

Knight-Jones, T. J. D., Edmond, K., Gubbins, S. and Paton, D. J. 2014. Veterinary and human vaccine evaluation methods. *Proc Biol Sci* 281(1784):20132839. doi:10.1098/rspb.2013.2839.

Koch, F., Thom, U., Albrecht, E., Weikard, R., Nolte, W., Kuhla, B. and Kuehn, C. 2019. Heat stress directly impairs gut integrity and recruits distinct immune cell populations into the bovine intestine. *Proc. Natl. Acad. Sci.* 116(21):10333-10338. doi:10.1073/PNAS.1820130116.

Kodjo, A., Villard, L., Bizet, C., Martel, J.-L., Sanchis, R., Borges, E., Gauthier, D., Oise Maurin, F. and Richard, Y. 1999. Pulsed-field gel electrophoresis is more efficient than ribotyping and random amplified polymorphic DNA analysis in discrimination of *Pasteurella haemolytica* strains. *J. Clin. Microbiol.* 37(2):380-385.

Lammers, B. P., vanKoot, J. W., Heinrichs, A. J. and Graves, R. E. 1996. The effect of plywood and polyethylene calf hutches on heat stress. *Appl. Eng. Agric.* 12(6):741-745.

Larson, R. L. and Step, D. L. 2012. Evidence-based effectiveness of vaccination against *Mannheimia haemolytica*, *Pasteurella multocida*, and *Histophilus somni* in feedlot cattle for mitigating the incidence and effect of bovine respiratory disease complex. *Vet. Clin. North Am. Food Anim. Pract.* 28(1):97-106, 106e1. doi:10.1016/j.cvfa.2011.12.005.

Le Grand, D., Solsona, M., Rosengarten, R. and Poumarat, F. 1996. Adaptive surface antigen variation in *Mycoplasma bovis* to the host immune response. *FEMS Microbiol. Lett.* 144(2-3):267-275. doi:10.1111/j.1574-6968.1996.tb08540.x.

Lima, S. F., de Souza Bicalho, M. L. S. and Bicalho, R. C. 2019. The *Bos taurus* maternal microbiome: role in determining the progeny early-life upper respiratory tract microbiome and health. *PLoS ONE* 14(3):e0208014. doi:10.1371/journal.pone.0208014.

Lima, S. F., Teixeira, A. G. V., Higgins, C. H., Lima, F. S. and Bicalho, R. C. 2016. The upper respiratory tract microbiome and its potential role in bovine respiratory disease and otitis media. *Sci. Rep.* 6:29050. doi:10.1038/srep29050.

Lippolis, K. D., Cooke, R. F., Schubach, K. M., Brandão, A. P., da Silva, L. G. T., Marques, R. S. and Bohnert, D. W. 2016. Altering the time of vaccination against respiratory pathogens to enhance antibody response and performance of feeder cattle. *J. Anim. Sci.* 94(9):3987-3995. doi:10.2527/jas.2016-0673.

Lombard, J., Urie, N., Garry, F., Godden, S., Quigley, J., Earleywine, T., McGuirk, S., Moore, D., Branan, M., Chamorro, M., Smith, G., Shively, C., Catherman, D., Haines, D., Heinrichs, A. J., James, R., Maas, J. and Sterner, K. 2020. Consensus recommendations on calf- and herd-level passive immunity in dairy calves in the United States. *J. Dairy Sci.* 103(8):7611-7624. doi:10.3168/jds.2019-17955.

Louie, A. P., Rowe, J. D., Love, W. J., Lehenbauer, T. W. and Aly, S. S. 2018. Effect of the environment on the risk of respiratory disease in preweaning dairy calves during summer months. *J. Dairy Sci.* 101(11):10230-10247. doi:10.3168/jds.2017-13716.

Love, W. J., Lehenbauer, T. W., Van Eenennaam, A. L., Drake, C. M., Kass, P. H., Farver, T. B. and Aly, S. S. 2016a. Sensitivity and specificity of on-farm scoring systems and nasal culture to detect bovine respiratory disease complex in preweaned dairy calves. *J. Vet. Diagn. Investig.* 28(2):119-128. doi:10.1177/1040638715626204.

Love, W. J., Lehenbauer, T. W., Karle, B. M., Hulbert, L. E., Anderson, R. J., Van Eenennaam, A. L., Farver, T. B. and Aly, S. S. 2016b. Survey of management practices related to bovine respiratory disease in preweaned calves on California dairies. *J. Dairy Sci.* 99(2):1483-1494. doi:10.3168/jds.2015-9394.

Love, W. J., Lehenbauer, T. W., Kass, P. H., Van Eenennaam, A. L. and Aly, S. S. 2014. Development of a novel clinical scoring system for on-farm diagnosis of bovine respiratory disease in pre-weaned dairy calves. *PeerJ* 2:e238. doi:10.7717/peerj.238.

Lysnyansky, I., Sachse, K., Rosenbusch, R., Levisohn, S. and Yogev, D. 2020. The vsp Locus of *Mycoplasma Bovis*: Gene Organization and Structural Features–PubMed. Accessed July 21, 2020. https://pubmed.ncbi.nlm.nih.gov/10482515/.

Lyte, M. 2004. Microbial endocrinology and infectious disease in the 21st century. *Trends Microbiol.* 12(1):14-20. doi:10.1016/j.tim.2003.11.004.

Lyte, M. 2014. The effect of stress on microbial growth. *Anim. Heal. Res. Rev.* 15(2):172-174. doi:10.1017/S146625231400019X.

Macaulay, A. S., Hahn, G. L., Clark, D. H. and Sisson, D. V. 1995. Comparison of calf housing types and tympanic temperature rhythms in Holstein calves. *J. Dairy Sci.* 78(4):856-862.

Mahan, S. M., Sobecki, B., Johnson, J., Oien, N. L., Meinert, T. R., Verhelle, S., Mattern, S. J., Bowersock, T. L. and Leyh, R. D. 2016. Efficacy of intranasal vaccination with a multivalent vaccine containing temperature-sensitive modified-live bovine herpesvirus type I for protection of seronegative and seropositive calves against respiratory disease. *J. Am. Vet. Med. Assoc.* 248(11):1280-1286. doi:10.2460/javma.248.11.1280.

Maier, G. U., Love, W. J., Karle, B. M., Dubrovsky, S. A., Williams, D. R., Champagne, J. D., Anderson, R. J., Rowe, J. D., Lehenbauer, T. W., Van Eenennaam, A. L. and Aly, S. S. 2019a. Management factors associated with bovine respiratory disease in preweaned calves on California dairies: the BRD 100 study. *J. Dairy Sci.* 102(8):7288-7305. doi:10.3168/jds.2018-14773.

Maier, G. U., Rowe, J. D., Lehenbauer, T. W., Karle, B. M., Williams, D. R., Champagne, J. D. and Aly, S. S. 2019b. Development of a clinical scoring system for bovine respiratory disease in weaned dairy calves. *J. Dairy Sci.* 102(8):7329-7344. doi:10.3168/jds.2018-15474.

Maier, G. U., Love, W. J., Karle, B. M., Dubrovsky, S. A., Williams, D. R., Champagne, J. D., Anderson, R. J., Rowe, J. D., Lehenbauer, T. W., Van Eenennaam, A. L. and Aly, S. S. 2020. A novel risk assessment tool for bovine respiratory disease in preweaned dairy calves. *J. Dairy Sci.* 103(10):9301-9317.

Mang, A. V., Buczinski, S., Booker, C. W. and Timsit, E. 2015. Evaluation of a computer-aided lung auscultation system for diagnosis of bovine respiratory disease in feedlot cattle. *J. Vet. Intern. Med.* 29(4):1112-1116. doi:10.1111/jvim.12657.

Mansfield, C. S., James, F. E. and Robertson, I. D. 2008. Development of a clinical severity index for dogs with acute pancreatitis. *J. Am. Vet. Med. Assoc.* 233(6):936-944. doi:10.2460/javma.233.6.936.

Mathew, A. G., Cissell, R. and Liamthong, S. 2007. Antibiotic resistance in bacteria associated with food animals: A United States perspective of livestock production. *Foodborne Pathog. Dis.* 4(2):115-133. doi:10.1089/fpd.2006.0066.

Maunsell, F., Brown, M. B., Powe, J., Ivey, J., Woolard, M., Love, W. and Simecka, J. W. 2012. Oral inoculation of young dairy calves with *Mycoplasma bovis* results in colonization of tonsils, development of otitis media and local immunity. *PLoS ONE* 7(9):e44523. doi:10.1371/journal.pone.0044523.

Maunsell, F. P. and Donovan, G. A. 2009. *Mycoplasma bovis* infections in young calves. *Vet. Clin. North Am. Food Anim. Pract.* 25(1):139-77, vii. doi:10.1016/j.cvfa.2008.10.011.

Maunsell, F. P., Donovan, G. A., Risco, C. and Brown, M. B. 2009. Field evaluation of a *Mycoplasma bovis* bacterin in young dairy calves. *Vaccine* 27(21):2781-2788. doi:10.1016/j.vaccine.2009.02.100.

Maunsell, F. P., Morin, D. E., Constable, P. D., Hurley, W. L., McCoy, G. C., Kakoma, I. and Isaacson, R. E. 1998. Effects of mastitis on the volume and composition of colostrum produced by Holstein cows. *J. Dairy Sci.* 81(5):1291-1299. doi:10.3168/jds.S0022-0302(98)75691-7.

Maunsell, F. P., Woolums, A. R., Francoz, D., Rosenbusch, R. F., Step, D. L., Wilson, D. J. and Janzen, E. D. 2011. *Mycoplasma bovis* infections in cattle. *J. Vet. Intern. Med.* 25(4):772-783. doi:10.1111/j.1939-1676.2011.0750.x.

McGill, J. L. and Sacco, R. E. 2020. The immunology of bovine respiratory disease: recent advancements. *Vet. Clin. North Am. Food Anim. Pract.* 36(2):333-348 doi:10.1016/j.cvfa.2020.03.002.

McGuirk, S. M. and Collins, M. 2004. Managing the production, storage, and delivery of colostrum. *Vet. Clin. North Am. Food Anim. Pract.* 20(3):593-603. doi:10.1016/j.cvfa.2004.06.005.

McMullen, C., Orsel, K., Alexander, T. W., van der Meer, F., Plastow, G. and Timsit, E. 2019. Comparison of the nasopharyngeal bacterial microbiota of beef calves raised without the use of antimicrobials between healthy calves and those diagnosed with bovine respiratory disease. *Vet. Microbiol.* 231:56-62. doi:10.1016/j.vetmic.2019.02.030.

Mitlöhner, F. M., Galyean, M. L. and McGlone, J. J. 2002. Shade effects on performance, carcass traits, physiology, and behavior of heat-stressed feedlot heifers. *J. Anim. Sci.* 80(8):2043-2050.

Morin, D. E., Nelson, S. V., Reid, E. D., Nagy, D. W., Dahl, G. E. and Constable, P. D. 2010. Effect of colostral volume, interval between calving and first milking, and photoperiod on colostral IgG concentrations in dairy cows. *J. Am. Vet. Med. Assoc.* 237(4):420-428. doi:10.2460/javma.237.4.420.

Muruganananthan, A., Shanthalingam, S., Batra, S. A., Alahan, S. and Srikumaran, S. 2018. Leukotoxin of *Bibersteinia trehalosi* contains a unique neutralizing epitope, and a non-neutralizing epitope shared with *Mannheimia haemolytica* leukotoxin. *Toxins (Basel)* 10(6). doi:10.3390/toxins10060220.

Nandi, S., Kumar, M., Manohar, M. and Chauhan, R. S. 2009. Bovine herpes virus infections in cattle. *Anim. Health Res. Rev.* 10(1):85-98. doi:10.1017/S1466252309990028.

Nandy, S., Parsons, S., Cryer, C., Underwood, M., Rashbrook, E., Carter, Y., Eldridge, S., Close, J., Skelton, D., Taylor, S., Feder, G. and Falls Prevention Pilot Steering Group 2004. Development and preliminary examination of the predictive validity of the Falls Risk Assessment Tool (FRAT) for use in primary care. *J. Publ. Heal.* 26(2):138-143. doi:10.1093/pubmed/fdh132.

Neupane, M., Kiser, J. N., Bovine Respiratory Disease Complex Coordinated Agricultural Project Research Team and Neibergs, H. L. 2018. Gene set enrichment analysis of SNP data in dairy and beef cattle with bovine respiratory disease. *Anim. Genet.* 49(6):527-538. doi:10.1111/age.12718.

Ng, T. F. F., Kondov, N. O., Deng, X., Van Eenennaam, A., Neibergs, H. L. and Delwart, E. 2015. A metagenomics and case-control study to identify viruses associated with bovine respiratory disease. *J. Virol.* 89(10):5340-5349. doi:10.1128/JVI.00064-15.

Nicola, I., Cerutti, F., Grego, E., Bertone, I., Gianella, P., D'Angelo, A., Peletto, S. and Bellino, C. 2017. Characterization of the upper and lower respiratory tract microbiota in *Piedmontese calves*. *Microbiome* 5(1):152. doi:10.1186/s40168-017-0372-5.

Nordlund, K. V. and Halbach, C. E. 2019. Calf barn design to optimize health and ease of management. *Vet. Clin. North Am. Food Anim. Pr.* 35(1):29-45.

Nosky, B., Biwer, J., Alkemade, S., Prunic, B., Milovanovic, A., Maletic, M., and Masic, A. 2017. Effect of a non-specific immune stimulant (amplimune™) on the health and production of light feedlot calves. *JDVAR* 6(3). doi:10.15406/jdvar.2017.06.00179.

O'Reilly, T. 1992. Enhancement of the effectiveness of antimicrobial therapy by muramyl peptide immunomodulators. *Clin. Infect. Dis.* 14(5):1100–1109. doi:10.1093/clinids/14.5.1100.

Oliver, S. P., Murinda, S. E. and Jayarao, B. M. 2011. Impact of antibiotic use in adult dairy cows on antimicrobial resistance of veterinary and human pathogens: a comprehensive review. *Foodborne Pathog. Dis.* 8(3):337–355. doi:10.1089/fpd.2010.0730.

Ollivett, T. L. and Buczinski, S. 2016. On-farm use of ultrasonography for bovine respiratory disease. *Vet. Clin. North Am. Food Anim. Pract.* 32(1):19–35. doi:10.1016/j.cvfa.2015.09.001.

Ollivett, T. L., Caswell, J. L., Nydam, D. V., Duffield, T., Leslie, K. E., Hewson, J. and Kelton, D. 2015. Thoracic ultrasonography and bronchoalveolar lavage fluid analysis in Holstein calves with subclinical lung lesions. *J. Vet. Intern. Med.* 29(6):1728–1734. doi:10.1111/jvim.13605.

Ollivett, T. L., Hewson, J., Schubotz, R. and Caswell, J. L. 2013. F-11 ultrasonographic progression of lung consolidation after experimental infection with *Mannheimia haemolytica* in Holstein Calves.

Palomares, R. A., Marley, S. M., Givens, M. D., Gallardo, R. A. and Brock, K. V. 2013. Bovine viral diarrhea virus fetal persistent infection after immunization with a contaminated modified-live virus vaccine. *Theriogenology* 79(8):1184–1195. doi:10.1016/j.theriogenology.2013.02.017.

Panciera, R. J. and Confer, A. W. 2010. Pathogenesis and pathology of bovine pneumonia. *Vet. Clin. North Am. Food Anim. Pract.* 26(2):191–214. doi:10.1016/j.cvfa.2010.04.001.

Pearce, S. C., Mani, V., Boddicker, R. L., Johnson, J. S., Weber, T. E., Ross, J. W., Rhoads, R. P., Baumgard, L. H. and Gabler, N. K. 2013a. Heat stress reduces intestinal barrier integrity and favors intestinal glucose transport in growing pigs. *PLoS ONE* 8(8):e70215. doi:10.1371/journal.pone.0070215.

Pearce, S. C., Mani, V., Weber, T. E., Rhoads, R. P., Patience, J. F., Baumgard, L. H. and Gabler, N. K. 2013b. Heat stress and reduced plane of nutrition decreases intestinal integrity and function in pigs. *J. Anim. Sci.* 91(11):5183–5193. doi:10.2527/jas.2013-6759.

Pearson, T. A. 2002. New tools for coronary risk assessment: what are their advantages and limitations? *Circulation* 105(7):886–892. doi:10.1161/hc0702.103727.

Perrett, T., Johnson, D. L., Song, J., Van De Pol, S., Dahlman, D. A., Rademacher, R. D., Hannon, S. J. and Booker, C. W. 2018. A retrospective analysis of feedlot morbidity and mortality outcomes in calves born to dams with known viral vaccination history. The Canadian veterinary journal = La revue veterinaire canadienne, 59(7):779–782.

Pithua, P., Aly, S. S., Haines, D. M., Champagne, J. D., Middleton, J. R. and Poock, S. E. 2013. Efficacy of feeding a lacteal-derived colostrum replacer or pooled maternal colostrum with a low IgG concentration for prevention of failure of passive transfer in dairy calves. *J. Am. Vet. Med. Assoc.* 243(2):277–282. doi:10.2460/javma.243.2.277.

Platt, R., Widel, P. W., Kesl, L. D. and Roth, J. A. 2009. Comparison of humoral and cellular immune responses to a pentavalent modified live virus vaccine in three age groups of calves with maternal antibodies, before and after BVDV type 2 challenge. *Vaccine* 27(33):4508–4519. doi:10.1016/j.vaccine.2009.05.012.

Pritchard, D. G., Carpenter, C. A., Morzaria, S. P., Harkness, J. W., Richards, M. S. and Brewer, J. I. 1981. Effect of air filtration on respiratory disease in intensively housed veal calves. *Vet. Rec.* 109(1):5–9. doi:10.1136/vr.109.1.5.

Purdy, C. W., Straus, D. C., Chirase, N., Parker, D. B., Ayers, J. R. and Hoover, M. D. 2002. Effects of aerosolized feedyard dust that contains natural endotoxins on adult sheep. *Am. J. Vet. Res.* 63(1):28–35. doi:10.2460/AJVR.2002.63.28.

Reef, V. B., Boy, M. G., Reid, C. F. and Elser, A. 1991. Comparison between diagnostic ultrasonography and radiography in the evaluation of horses and cattle with thoracic disease: 56 cases (1984–1985). *J. Am. Vet. Med. Assoc.* 198(12):2112–2118.

Ribeiro, M. G., Risseti, R. M., Bolaños, C. A. D. D., Caffaro, K. A., de Morais, A. C. B. B., Lara, G. H. B. B., Zamprogna, T. O., Paes, A. C., Listoni, F. J. P. P. and Franco, M. M. J. J. 2015. *Trueperella pyogenes* multispecies infections in domestic animals: a retrospective study of 144 cases (2002 to 2012). *Vet. Q.* 35(2):82–87. doi:10.1080/01652176.2015.1022667.

Rice, J. A., Carrasco-Medina, L., Hodgins, D. C. and Shewen, P. E. 2007. Mannheimia haemolytica and bovine respiratory disease. *Anim Health Res Rev.* 8(2):117–128. doi:10.1017/S1466252307001375. PMID: 18218156.

Richeson, J. T. and Falkner, T. R. 2020. Bovine respiratory disease vaccination: what is the effect of timing? *Vet. Clin. North Am. Food Anim. Pract.* 36(2):473–485. doi:10.1016/j.cvfa.2020.03.013.

Ridpath, J. F., Fulton, R. W., Bauermann, F. V., Falkenberg, S. M., Welch, J. and Confer, A. W. 2020. Sequential exposure to bovine viral diarrhea virus and bovine coronavirus results in increased respiratory disease lesions: clinical, immunologic, pathologic, and immunohistochemical findings. *J. Vet. Diagn. Investig.* 32(4):513–526. doi:10.1177/1040638720918561.

Roland, L., Drillich, M., Klein-Jöbstl, D. and Iwersen, M. 2016. Invited review: influence of climatic conditions on the development, performance, and health of calves. *J. Dairy Sci.* 99(4):2438–2452. doi:10.3168/jds.2015-9901.

Romanowski, R., Culbert, R., Alkemade, S., Medellin-Peña, M. J., Bugarski, D., Milovanovic, A., Nesic, S. and Masic, A. 2017. *Mycobacterium* cell wall fraction immunostimulant ("amplimune") efficacy in the reduction of the severity of etec induced diarrhea in neonatal calves. *Acta Vet. Brno* 67(2):222–237. doi:10.1515/acve-2017-0019.

Scholthof, K. B. G. 2007. The disease triangle: pathogens, the environment and society. *Nat. Rev. Microbiol.* 5(2):152–156. doi:10.1038/nrmicro1596.

Snowder, G. D., Van Vleck, L. D., Cundiff, L. V. and Bennett, G. L. 2006. Bovine respiratory disease in feedlot cattle: environmental, genetic, and economic factors. *J. Anim. Sci.* 84(8):1999–2008. doi:10.2527/jas.2006-046.

Speer, B. S., Shoemaker, N. B. and Salyers, A. A. 1992. Bacterial resistance to tetracycline: mechanisms, transfer, and clinical significance. *Clin Microbiol Rev* 5(4):387–399.

Srikumaran, S., Kelling, C. L. and Ambagala, A. 1996. Immune evasion by pathogens of bovine respiratory disease complex. *Anim. Heal. Res. Rev.* 8:215–229. doi:10.1017/S1466252307001326.

Stevens, E. T., Brown, M. S., Burdett, W. W., Bolton, M. W., Nordstrom, S. T. and Chase, C. C. L. (2010). Efficacy of a non-adjuvanted, modified-live virus vaccine in calves with maternal antibodies against a virulent bovine viral diarrhea virus type 2a challenge seven months following vaccination. The Bovine Practitioner, 45(1):23-31. https://doi.org/10.21423/bovine-vol45no1p23-31.

Stott, G. H., Marx, D. B., Menefee, B. E. and Nightengale, G. T. 1979. Colostral immunoglobulin transfer in calves II. The rate of absorption. *J. Dairy Sci.* 62(11):1766-1773. doi:10.3168/jds.S0022-0302(79)83495-5.

Stroebel, C., Alexander, T., Workentine, M. L. and Timsit, E. 2018. Effects of transportation to and co-mingling at an auction market on nasopharyngeal and tracheal bacterial communities of recently weaned beef cattle. *Vet. Microbiol.* 223:126-133. doi:10.1016/j.vetmic.2018.08.007.

Svensson, C., Lundborg, K., Emanuelson, U. and Olsson, S. O. 2003. Morbidity in Swedish dairy calves from birth to 90 days of age and individual calf-level risk factors for infectious diseases. *Prev. Vet. Med.* 58(3-4):179-197. doi:10.1016/S0167-5877(03)00046-1.

Theurer, M. E., Larson, R. L. and White, B. J. 2015. Systematic review and meta-analysis of the effectiveness of commercially available vaccines against bovine herpesvirus, Bovine viral diarrhea virus, Bovine respiratory syncytial virus, and parainfluenza type 3 virus for mitigation of bovine respiratory disease complex in cattle. *J. Am. Vet. Med. Assoc.* 246(1):126-142. doi:10.2460/javma.246.1.126.

Thomas, C. B., Van Ess, P., Wolfgram, L. J., Riebe, J., Sharp, P. and Schultz, R. D. 1991. Adherence to bovine neutrophils and suppression of neutrophil chemiluminescence by *Mycoplasma bovis*. *Vet. Immunol. Immunopathol.* 27(4):365-381. doi:10.1016/0165-2427(91)90032-8.

Thomas, L. H., Stott, E. J., Collins, A. P., Jebbett, N. J. and Stark, A. J. 1977. Evaluation of respiratory disease in calves: comparison of disease response to different viruses. *Res. Vet. Sci.* 23(2):157-164. doi:10.1016/s0034-5288(18)33148-5.

Timsit, E., Holman, D. B., Hallewell, J. and Alexander, T. W. 2016a. The nasopharyngeal microbiota in feedlot cattle and its role in respiratory health. *Anim. Front.* 6(2):44-50. doi:10.2527/af.2016-0022.

Timsit, E., Workentine, M., Schryvers, A. B., Holman, D. B., van der Meer, F. and Alexander, T. W. 2016b. Evolution of the nasopharyngeal microbiota of beef cattle from weaning to 40 days after arrival at a feedlot. *Vet. Microbiol.* 187:75-81. doi:10.1016/j.vetmic.2016.03.020.

Timsit, E., Workentine, M., van der Meer, F. and Alexander, T. 2018. Distinct bacterial metacommunities inhabit the upper and lower respiratory tracts of healthy feedlot cattle and those diagnosed with bronchopneumonia. *Vet. Microbiol.* 221:105-113. doi:10.1016/j.vetmic.2018.06.007.

Traub, S., von Aulock, S., Hartung, T. and Hermann, C. 2006. MDP and other muropeptides–direct and synergistic effects on the immune system. *J. Endotoxin Res.* 12(2):69-85. doi:10.1179/096805106X89044.

Urie, N. J., Lombard, J. E., Shivley, C. B., Kopral, C. A., Adams, A. E., Earleywine, T. J. and Olson, J. D. 2020. Since January 2020 Elsevier has created a COVID-19 Resource Centre with free information in English and Mandarin on the novel coronavirus COVID-19. The COVID-19 resource centre is hosted on Elsevier Connect, the Company's Public News and Information.

Urie, N. J. N., Lombard, J. E., Shivley, C. B. C., Kopral, C. A., Adams, A. E., Earleywine, T. J. T., Olson, J. D. J. and Garry, F. B. 2018. Preweaned heifer management on US dairy operations: Part V. Factors associated with morbidity and mortality in preweaned dairy heifer calves. *J. Dairy Sci.* 101(10):9229-9244. doi:10.3168/jds.2017-14019.

USDA 2010. Beef 2007–08 Part IV: Reference of beef cow-calf management practices in the United States, 2007–2008. USDA:APHIS:VS, CEAH. Fort Collins, CO.

USDA 2011. How to do risk assessments and develop management plans for Johne's Disease. Fourth Ed. https://www.oregon.gov/ODA/shared/Documents/Publications/AnimalHealth/HowToAssessmentsJD.pdf.

USDA 2016. Dairy 2014 Dairy Cattle Management Practices in the United States, 2014. USDA, Animal and Plant Health Information Service, Veterinary Services, National Animal Health Monitoring System, Fort Collins, CO.

USDA 2018. Dairy 2014. *Health and Management Practices on U.S. Dairy Operations*, USDA-APHIS-VS-CEAH-NAHMS. Fort Collins, CO #696.0218.

Vaarten, J. 2012. Clinical impact of antimicrobial resistance in animals. *Rev. Sci. Tech.* 31(1):221–229.

Van Donkersgoed, J., Ribble, C. S., Boyer, L. G. and Townsend, H. G. 1993. Epidemiological study of enzootic pneumonia in dairy calves in Saskatchewan. *Can. J. Vet. Res.* 57(4):247–254.

Van Driessche, L., Valgaeren, B. R., Gille, L., Boyen, F., Ducatelle, R., Haesebrouck, F., Deprez, P. and Pardon, B. 2017. A deep nasopharyngeal swab Versus nonendoscopic bronchoalveolar lavage for isolation of bacterial pathogens from preweaned calves with respiratory disease. *J. Vet. Intern. Med.* 31(3):946–953. doi:10.1111/jvim.14668.

Van Eenennaam, A., Neibergs, H., Seabury, C., Taylor, J., Wang, Z., Scraggs, E., Schnabel, R. D., Decker, J., Wojtowicz, A., Aly, S., Davis, J., Blanchard, P., Crossley, B., Rossitto, P., Lehenbauer, T., Hagevoort, R., Chavez, E., Neibergs, J. S. and Womack, J. E. 2014. Results of the BRD CAP project: progress toward identifying genetic markers associated with BRD susceptibility. *Anim. Heal. Res. Rev.* 15(2):157–160. doi:10.1017/S1466252314000231.

Vangeel, I., Ioannou, F., Riegler, L., Salt, J. S. and Harmeyer, S. S. 2009. Efficacy of an intranasal modified live bovine respiratory syncytial virus and temperature-sensitive parainfluenza type 3 virus vaccine in 3-week-old calves experimentally challenged with PI3V. *Vet. J.* 179(1):101–108. doi:10.1016/j.tvjl.2007.08.008.

Veit, H. P. and Farrell, R. L. 1978. The anatomy and physiology of the bovine respiratory system relating to pulmonary disease. *Cornell Vet.* 68(4):555–581.

Waldner, C. L., Parker, S. and Campbell, J. R. 2019. Vaccine usage in western Canadian cow-calf herds. *Can. Vet. J.* 60(4):414–422.

Wathes, C. M., Jones, C. D. and Webster, A. J. 1983. Ventilation, air hygiene and animal health. *Vet. Rec.* 113(24):554–559.

Watts, J. L. and Sweeney, M. T. 2010. Antimicrobial resistance in bovine respiratory disease pathogens: measures, trends, and impact on efficacy. *Vet. Clin. Food Anim.* 26(1):79–88. doi:10.1016/j.cvfa.2009.10.009.

Weaver, D. M., Tyler, J. W., VanMetre, D. C., Hostetler, D. E. and Barrington, G. M. 2000. Passive transfer of colostral immunoglobulins in calves. *J. Vet. Intern. Med.* 14(6):569–577. doi:10.1892/0891-6640(2000)014<0569:ptocii>2.3.co;2.

Weinberg, G. A. and Szilagyi, P. G. 2010. Vaccine epidemiology: efficacy, effectiveness, and the translational research roadmap. *J. Infect. Dis.* 201(11):1607–1610. doi:10.1086/652404.

Welsh, R. D., Dye, L. B., Payton, M. E. and Confer, A. W. 2004. Isolation and antimicrobial susceptibilities of bacterial pathogens from bovine pneumonia: 1994–2002. *J. Vet. Diagn. Investig.* 16(5):426–431. doi:10.1177/104063870401600510.

Williams, D. R., Pithua, P., Garcia, A., Champagne, J., Haines, D. M. and Aly, S. S. 2014. Effect of three colostrum diets on passive transfer of immunity and preweaning health in calves on a California dairy following colostrum management training. *Vet. Med. Int.* 2014:698741. doi:10.1155/2014/698741.

Windeyer, M. C. and Gamsjäger, L. 2019. Vaccinating calves in the face of maternal antibodies: challenges and opportunities. *Vet. Clin. North Am. Food Anim. Pract.* 35(3):557-573. doi:10.1016/j.cvfa.2019.07.004.

Windeyer, M. C., Leslie, K. E., Godden, S. M., Hodgins, D. C., Lissemore, K. D. and LeBlanc, S. J. 2012. The effects of viral vaccination of dairy heifer calves on the incidence of respiratory disease, mortality, and growth. *J. Dairy Sci.* 95(11):6731-6739. doi:10.3168/jds.2012-5828.

Windeyer, M. C., Leslie, K. E., Godden, S. M., Hodgins, D. C., Lissemore, K. D. and LeBlanc, S. J. 2014. Factors associated with morbidity, mortality, and growth of dairy heifer calves up to 3 months of age. *Prev. Vet. Med.* 113(2):231-240. doi:10.1016/j.prevetmed.2013.10.019.

Windeyer, M. C., Timsit, E. and Barkema, H. 2017. Bovine respiratory disease in pre-weaned dairy calves: are current preventative strategies good enough? *Vet. J.* 224:16-17. doi:10.1016/j.tvjl.2017.05.003.

Woolhouse, M. E. J., Haydon, D. T. and Bundy, D. A. P. 1997. The design of veterinary vaccination programmes. *Vet. J.* 153(1):41-47. doi:10.1016/S1090-0233(97)80007-X.

Woolums, A. R., Berghaus, R. D., Berghaus, L. J., Ellis, R. W., Pence, M. E., Saliki, J. T., Hurley, K. A. E., Galland, K. L., Burdett, W. W., Nordstrom, S. T. and Hurley, D. J. 2013. Effect of calf age and administration route of initial multivalent modified-live virus vaccine on humoral and cell-mediated immune responses following subsequent administration of a booster vaccination at weaning in beef calves. *Am. J. Vet. Res.* 74(2):343-354. doi:10.2460/ajvr.74.2.343.

Woolums, A. R. and Step, D. L. 2020. Bovine respiratory disease: what's new? *Vet. Clin. North Am. Food Anim. Pract.* 36(2):xv-xvi. doi:10.1016/j.cvfa.2020.04.001.

Zachary, J. 2017 *Pathologic Basis of Veterinary Disease* (6th ed. Zachary, J. F., Ed.). Elsevier, St. Louis Missouri 63043.

Zhang, M., Hill, J. E., Fernando, C., Alexander, T. W., Timsit, E., van der Meer, F. and Huang, Y. 2019. Respiratory viruses identified in western Canadian beef cattle by metagenomic sequencing and their association with bovine respiratory disease. *Transbound. Emerg. Dis.* 66(3):1379-1386. doi:10.1111/tbed.13172.

Chapter 3

The use of probiotics as supplements for ruminants

Frédérique Chaucheyras-Durand and Lysiane Dunière, Lallemand Animal Nutrition and Université Clermont Auvergne, INRAE, UMR 454 MEDIS, France

1 Introduction

Good rumen function is one of the most important characteristics in cattle as it impacts their health and productivity and, thus, helps to ensure sufficient food production for a growing human population. In its life, a ruminant faces several periods of stress, from gestation and calving to the demands of high production of milk or meat. Current intensive farming practices are known to challenge digestive microbiota and put the animal at risk of developing metabolic disorders. Much research has been devoted recently for investigating the impact of some of these life cycle stages on rumen function and health. This research has shown that the rumen microbiota and the rumen epithelial wall play a key role to maintain optimal function during these stages.

http://dx.doi.org/10.19103/AS.2020.0067.26

This chapter begins by reviewing critical periods in the ruminant lifecycle as targets for probiotics. It then looks at definitions of probiotics, delivery mechanisms and regulation. The rest of the chapter summarizes and assesses the range of research on the benefits and modes of action of probiotics, starting with their potential in young ruminants. It then considers the role of probiotics in adult ruminants in the following areas: feed efficiency, methane production, pathogen control and supporting the immune system.

2 Critical periods in the ruminant lifecycle as targets for probiotics

Calving is an extremely stressful event for the neonate ruminant, especially from the microbiological point of view. Colonization of the gastrointestinal tract (GIT) starts during and after the birth through direct contact with the mother and farm environment (Yeoman et al., 2018). Members of typical ruminal populations, such as methanogens, fibrolytic bacteria or Proteobacteria, have been detected in the rumen of calves less than 20 minutes after birth (Guzman et al., 2015) and are metabolically active (Guzman et al., 2016). Neonate calf physiology (i.e. esophageal groove closure reflex) prevents milk from entering the rumen compartment during the pre-weaning phase. However, the immaturity of the groove and consumption of small amounts of solids contribute to rumen microbial inoculation and a slow and immature development of rumen microbiota and physiology before weaning (Meale et al., 2017). Colonization of the rumen starts with aerobic and facultative anaerobic taxa which are gradually replaced by strict anaerobic microorganisms (Rey et al., 2014). Impairment in rumen development will affect nutrient digestion and impact the further growth and performance of the animal (Steele et al., 2016).

Digestive issues are one of the most common reasons for disease-related deaths among dairy calves and heifers. An average mortality rate as high as 56.4% has been observed in the pre-weaning period for heifers in the North American dairy industry (USDA, 2014a). Most calf diarrhea problems occur within the first 3 weeks of life (McGuirk, 2008). In a 2-year study in Estonia, metabolic and digestive disorders were identified as the main reason for on-farm mortality in calves under 1 month of age (43%), between 1-5 months (29.5%) and between 6-19 months of age (32.3%) (Mõtus et al., 2018). Poor production linked to digestive disorders is the identified cause of removing pre-weaned heifers in 21.1% of cases in the United States (USDA, 2014a).

At birth, calf immunity depends on colostrum and milk consumption which allows transfer of immunoglobulins (Ig) and neutrophils, as well as macrophages known to secrete immune-related components such as cytokines or antimicrobial peptides and proteins (Stelwagen et al., 2009). IgG is the main immunoglobulin in colostrum and its concentration is used to characterize colostrum quality (Johnsen et al., 2019). Low quality or limited consumption of colostrum increases mortality and decreases calf growth (McGuirk and Collins, 2004), while feeding high-quality colostrum soon after birth increases calf weaning weight and BW gain (Priestley et al., 2013). These results highlight the link between diet and immune function. Colostral immunity is recognized as an essential part of GIT disease management in dairy calves (McGuirk, 2008). The proportion of *E. coli* attached to small intestinal tissue was significantly higher for calves deprived of colostrum when compared to colostrum-fed animals, whereas the proportion of *Bifidobacterium* was lower (Malmuthuge et al., 2015). In newborn lambs, the gut microbiome was speculated to provide critical signals involved in the establishment of a functional immune response, for example, in maintaining the development and function of lymphoid follicles in ileal Peyer patches (Reynolds and Morris, 1984).

Weaning usually occurs around 6-8 weeks of age for dairy calves and represents one of the most dramatic changes in rumen development, leading to important modifications in intestinal mass, immunity and metabolism (Baldwin et al., 2004). It is recognized that a badly managed weaning stage will often be associated with distress, depressed growth and diarrhea (Roth et al., 2009). At weaning, milk is abruptly or gradually removed from the calf diet and replaced by solid feed entering the rumen compartment. At that time, overall microbial activity increases, as shown by the growth of several members of rumen microbiota, the increase in ruminal enzyme production and resulting fermentative end-products (Jiao et al., 2015). Weaning leads to a strong and rapid physiological evolution of the rumen (e.g. in papillae development, volume expansion, increased rumen wall thickness) (reviewed in Meale et al., 2017). A decrease in microbial diversity has been observed immediately after weaning, probably due to drastic dietary change from the transition to a feed-based diet, extensive fermentation and fluctuating pH in the rumen compartment (Meale et al., 2016). Rumen development in calves is also affected by the physical form and chemical composition of starter diets (Khan et al., 2016).

The mucosal epithelium is considered to be a barrier between the GIT and the host. Epithelial cells play a key role in recognizing the rumen microbiome, pathogens and chemicals present in digesta, thus influencing development of the mucosal immune system (Malmuthuge et al., 2012). Toll-Like Receptors (TLRs) found on mucosal epithelium all along the GIT are down-regulated during continuous exposure to microbial ligands coming from rumen microbiota in order to limit a chronic and unnecessary inflammatory response. Gene expression of epithelial expressed molecules such as β-defensin (antimicrobial peptide) and PGLYRP1 (peptidoglycan recognition protein) are repressed prior to weaning but increase afterwards. Malmuthuge et al. (2012) suggested that TLRs drive primary innate immunity up to weaning but that their role decreases in favor of other innate immune mechanisms as the animal ages. Bush et al. (2019) conducted a comprehensive transcriptomic study of the whole GIT of ruminants from birth to adulthood and observed a strong immune transcriptomic signature in rumen evolving with animal age.

Research focusing on the developing rumen has demonstrated that it may be possible to regulate microbial community development by controlling feeding management early in life, with subsequent effects later in the animal's life (Abecia et al., 2013, 2014; Yáñez-Ruiz et al., 2015). Alteration of the diet of goat kids to reduce methane emissions was shown to modify bacterial and archaeal rumen populations up to 4 months after treatment; there was also a maternal effect since modifications of bacterial groups and metabolite profiles were also associated with kids from treated mothers (Abecia et al., 2018). These results also highlight the potential long-term effect of maternal imprinting on rumen microbiota. A meta-analysis by Soberon and Van Amburgh (2013) observed that heifers fed milk ab-libitum during the pre-weaning period showed a greater milk yield in the first lactation than calves with a restricted milk diet during pre-weaning. Feeding a higher level of nutrients during the pre-weaning period will affect cell functions involved in the morphological and physiological development of the mammary gland and increase milk yield (Hare et al., 2019). These data suggest that dietary interventions in early life, although not specifically targeting the rumen, can have an impact on later productivity of the animal.

During their life ruminants will need to be transported from their place of birth and mixed with other animals in places such as another farm, feedlots or market auctions. In 2013, 28.6% of US dairy farms introduced new animals into their herd (USDA, 2014b). This period is associated with several stresses such as handling and transportation, feed or water privation, mixing with other animals or high temperature variations. It is well known that stress negatively affects the immune system (Blecha et al., 1984) which could be very detrimental at a time when the animal is more likely to be exposed to infectious agents as a result of commingling with other animals. Stress will impair cattle performance and health with a decreased feed intake being commonly observed (Hutcheson and Cole, 1986; Silanikove, 2000). Ashenafi et al. (2018) reviewed the effect of stresses associated with transportation over long distances and noted alteration in metabolism, immune competence and behavior, failures in reproduction and increases in morbidity and mortality.

Beef cattle during the fattening period or dairy cows during the lactating period are commonly fed high-grain diets in order to meet energy requirements. The animals are supplied with large amounts of readily fermentable starch or sugars, known to alter rumen microbial communities and functions (Petri et al., 2013a,b, 2018). As rumen pH falls, lactate producers may outnumber lactate utilizers leading to changes in the structure of rumen microbiota (Russell and

Wilson, 1996). The abundance of fibrolytic species *Ruminococcus* sp. and *Fibrobacter succinogenes* was shown to decrease in rumen of cows switching from a forage to high-grain diet. In contrast, lactate utilizers *Selenomonas ruminantium* and *Megasphaera elsdenii* increased at that time (Petri et al., 2013b). *Lactobacillus* and *Streptococcus* were observed to appear in the rumen of clinically acidotic cows, probably reflecting the tolerance of these species to low pH and their ability to proliferate on an excess of fermentable carbohydrate (Petri et al., 2013a). Low rumen pH for a prolonged period can negatively affect feed intake, microbial metabolism and nutrient degradation, and leads to physiological disorders such as acidosis, inflammation, laminitis, diarrhea and milk fat depression (Kleen et al., 2003; Villot et al., 2018).

This microbial dysbiosis may trigger the release of potential harmful molecules which impact animal health. Acidosis has been associated with release of lipopolysaccharide (LPS) from Gram-negative bacteria into the rumen and the hindgut (Khafipour et al., 2009; Khiaosa-Ard and Zebeli, 2018). LPS translocation has been observed from the rumen into the interior circulation which can trigger an inflammatory response with an increase in acute phase protein concentration in peripheral blood (reviewed in Plaizier et al., 2012). Histamine (and other biogenic amines) are produced by some rumen bacteria under low pH conditions (Silberberg et al., 2013; Wang et al., 2013), leading to an increase in rumen epithelial cell inflammatory response (Sun et al., 2017). Finally, a more global systemic inflammation response may be triggered by an activation of the immune system in the bloodstream or tissues, when a significant amount of nutrients and energy is diverted from maintenance of general body homeostasis and production, which ultimately results in poor animal performance with a significant economic impact (Zebeli et al., 2015). Higher expression of TLR2 and TLR4 was identified in the rumen papillaes of acidosis-resistant vs. sensitive steers indicating greater host innate immune response. The authors suggested that an increased expression of TLRs may protect the rumen epithelium from subacute ruminal acidosis (SARA) damage by stimulating the barrier function of the rumen in resistant steers (Chen et al., 2012). It has been shown that leukocytes present in saliva can migrate back and forth from the rumen cavity, starting a cross-talk with the lymphoid tissues in the oral cavity and cytokines or other mediators released by forestomach walls. In a field trial on 128 cows, it has been shown that concentrations of total Ig and IgM in rumen fluid were regulated by rumen pH levels, as well as by volatile fatty acids (VFA) concentrations. These findings suggest control by the innate immune system over metabolic activities in bovine forestomachs (Trevisi et al., 2018).

Epimural microbiota is attached to the ruminal epithelium and is in close contact with epithelial cells involved in inflammatory response modulation. Epimural microbial diversity is distinct from microbiota observed either in liquid or solid phases in the rumen (Sadet et al., 2007). Diversity of the rumen epimural population is strongly affected with a shift from *Firmicutes* to *Proteobacteria* as concentrates are replaced by hay, although the expression of genes targeting intracellular pattern recognition receptor (TLR), barrier function, pH regulation, and nutrient uptake of rumen epithelium remain stable (Petri et al., 2018). Chen et al. (2011) observed the differences in epimural microbiota under high-hay or high-grain diet with *Treponema* sp., *Ruminobacter* sp., and *Lachnospiraceae* observed only with the high-grain diet. These results contradicted previous published data and the authors suggested that the host had a role in regulating microbial diversity and density and in the response to rumen environmental changes.

At the start of the lactation period, dairy cows are often unable to reach their energy requirements and need to mobilize body reserves, despite the alteration in diet from high-fiber to high-energy content. Rumen function should thus be maintained at its best to get the best reproductive performance and milk production. At that time, the rumen undergoes drastic changes impacting rumen microbiota, fermentation profile and epithelial permeability (Bach et al., 2018; Minuti et al., 2015). Zhu et al. (2018) observed changes in bacterial and archaeal communities which could be linked to short-chain fatty acids profiles over the transition period in primiparous dairy cows. The relative abundance of *Bacteroidetes* decreased and *Proteobacteria* increased after calving. Significant shifts in methanogenic community composition over the transition period were observed mainly for *Methanosphaera* and *Methanomassiliicoccus*, although the pattern varied across genera.

Rumen microbial composition appears to be linked to milk production. The ratio of *Firmicutes* to *Bacteroidetes* in rumen has been negatively correlated to feed efficiency (FE) and changes in milk fat yield in Holstein cattle (Jami et al., 2014). A study to determine the associations between the rumen microbiota and feed efficiency was conducted in a Holstein cattle using whole metagenome sequencing. A larger relative abundance of *Bacteroidetes* and *Prevotella* was observed in more efficient cows together with a lower relative abundance of *Firmicutes* and some members of the archaeal population (Delgado et al., 2019).

The transition period (from 3 weeks before to 3 weeks after calving) is linked with drastic effects on cow health and has been associated with a reduction of immune competence, a negative energy balance, hypocalcemia, an over-systemic inflammatory response (even in the absence of signs of microbial infection) and a state of oxidative stress (Trevisi and Minuti, 2018). The expression of genes involved in immune response modulation in rumen epithelium was observed to vary widely from two weeks before calving to 3 weeks post-partum (Bach et al., 2018). The inflammatory response is considered to be directly correlated with the increase in the release of ruminal endotoxins due to high-concentrate diets during lactation (Abaker et al., 2017). Ingvartsen et al. (2003) reviewed the diseases associated with high-yielding cows and found that imbalanced immune status and diets lead to metabolic disorders and reproduction issues, mastitis and increased risk of ketosis and lameness.

Pitta et al. (2018) reviewed recent studies on microbes and dietary interactions to enhance the productivity of dairy cows. They have identified a number of strategies to manipulate rumen microbial processes (Fig. 1). The potential action of probiotics can be considered in different areas such as fiber digestion, protein supply and microbial growth, bio-hydrogenation and methane production. The ruminant is influenced by its microbiota and immune system but also by dietary interventions and environmental stresses. The manipulation of gastrointestinal microbiota in maintaining animal gut health, through understanding diversity, stability, metabolites and crosstalk with the epithelium and the underlying immune system, still needs further research (Gaggìa et al., 2010).

3 Definitions, delivery mechanisms and regulation

3.1 Definitions

The term probiotic comes from the Latin 'pro' ('for') and the Greek 'bios' ('life') and was firstly suggested in the 1960s in contrast to antibiotic ('against life') to define substances produced by protozoa that are able to support the growth of other microorganisms (Morelli and Capurso, 2012). In the late 1980s, Dr R. Fuller proposed a definition for probiotic as 'a live microbial feed supplement which beneficially affects the host animal by improving its intestinal

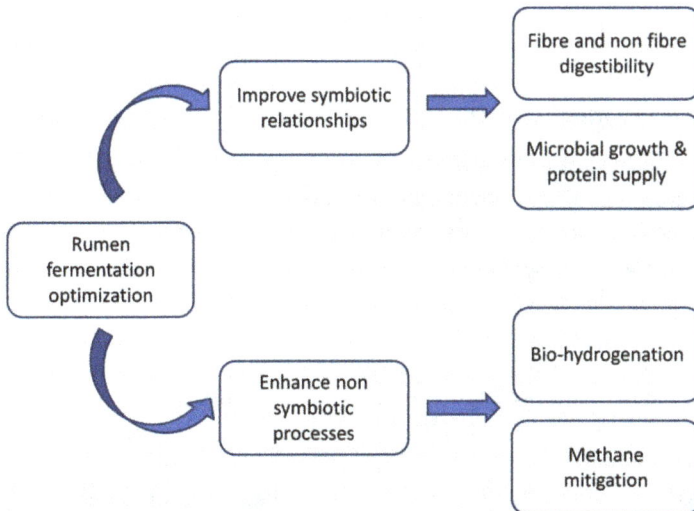

Figure 1 Strategies to manipulate rumen microbial processes to enhance rumen functions. Source: adapted from Pitta et al. (2018).

balance'. At that time, probiotics were only focused on livestock animals. With the expansion of use in humans, the Food and Agriculture Organization of the United Nations (FAO)/World Health Organization (WHO) proposed a slightly different definition in 2001: 'live microorganisms which when administered in adequate amounts confer a health benefit on the host'. This definition was reviewed in 2013 by a panel of scientific experts, the International Scientific Association for Probiotics and Prebiotics, who agreed that the FAO/WHO definition for probiotics was still relevant. In this chapter, we will only consider live microorganisms and not dead or inactivated microorganisms, such as yeast culture or microbial fractions, which are not probiotics as defined above.

We need to underline that, in the case of ruminants, the term 'probiotic' is in fact not commonly used. Indeed the strict definition is not totally appropriate for several reasons:

- the main digestive target is the rumen and not the intestine.
- the focus is mainly nutritional effects rather than health benefits.
- live microorganisms are being generally incorporated in feed so have been considered as feed supplements.
- in European Union (EU) and other countries, these microorganisms fall into the feed additive category (zootechnical additives) (EC Regulation 1831/2003).

In most countries, commercialized products containing live microorganisms are considered as feed additives (FA) or direct-fed microbials (DFM) and these terms will be used throughout this chapter rather than 'probiotic'.

3.2 Delivery mechanisms

In ruminants, microbial FA are delivered through the diet, either with a mineral/vitamin premix, in milk powder or included in pelleted concentrate. As the concentration of active cells per gram of pure additive is high, inclusion in feed is generally at a low percentage which ensures a precise amount of additive per animal. It is important to assess the compatibility of microbial FA with some minerals (i.e. copper) which may exhibit some toxicity toward the live cells. It is also important to ensure that the pelleting process, which involves conditions such as high temperature, high pressure and humidity, will not damage microbial cells. Sullivan and Bradford (2011) have raised this concern and compared different commercial active dry yeasts (ADY) containing highly concentrated live yeast cells fed to dairy cattle. Tested products failed to consistently meet product claims and viability was significantly diminished during storage at 40°C for 3 weeks. Loss in viability at elevated temperatures can be reduced when

ADY products are diluted with a premix containing vitamin trace minerals due to the antioxidant role of vitamins A and E and micro minerals such as selenium. FA producers need to ensure the right formulations to address these problems. In the case of inclusion into pelleted concentrates, some companies have proposed technologies such as micro-encapsulated beads to protect microbial FA against harsh feed manufacturing conditions and ensure optimal stability of live microorganisms up to the point of consumption. In case of combination with other feed supplements, such as essential oils, plant extracts or antibiotics, it is important to ensure that the live microorganisms will not be negatively affected by these compounds which are known to exert antimicrobial effects. A particular situation is feedlot cattle supplementation. In large-scale fattening operations, it is possible to feed the animals with liquid supplements containing DFM, primarily to control pathogens in the gut (discussed later in this chapter). In this case, DFM (bacteria) are delivered in refrigerated containers and fed daily to the herd, avoiding problems about stability in feed.

In many species including humans, the concept of probiotics is based on a daily distribution of live microorganisms as these microorganisms do not need to colonize the digestive tract to be active. In ruminants, microbial FA are also administered daily through feed. It has been shown that a repeated distribution of a live yeast product ensured a stable concentration of cells in the rumen of lambs (Durand-Chaucheyras et al., 1998). However, when supplementation was stopped, a decrease in live cell concentration was measured after 24h and a total loss from the rumen was observed after a few days, with some recovery from the feces of animals up to several days after the last incorporation of the yeast product in the diet.

Some yeasts, such as *Saccharomyces boulardii*, are recognized for their ability to survive in the gut and to reach the lower gut compartment alive, justifying its current application in pre-weaned ruminants in North America to promote intestinal health. Probiotic bacteria are mainly targeted at lower gut health. However, there is limited information about their potential survival in this part of the gut. This is, in part, due to the lack of detection methods allowing a specific detection/quantification of live/viable probiotic bacteria within the commensal bacterial population in the gut, since the latter often contains the same species as those used in the probiotic (Fomenky et al., 2017). However it is clear that the digestive environment of the ruminant provides hostile conditions for survival of probiotics. Research is needed to develop innovative formulations which could help deliver active microorganisms in the lower gut. Microencapsulation is an emerging technology used to protect probiotics against adverse environmental conditions. Different microencapsulation techniques can be used with materials such as alginate, chitosan, carrageenan, gums, gelatin, whey protein, starch and the use of compression coating (Riaz and Masud, 2013). It is important to find an optimal balance between capsule

Table 1 Yeast and bacteria species used as probiotics in ruminants

Yeasts	Bacteria
Saccharomyces cerevisiae	Lactobacillus acidophilus/crispatus
Saccharomyces boulardii	Lactobacillus rhamnosus
	Enterococcus faecium
	Bacillus subtilis
	Bacillus licheniformis
	Propionibacterium freudenreichii
	Propionibacterium acidipropionici
	Megasphaera elsdenii

characteristics, protection of the probiotic, added cost and ease of use. Table 1 summarizes microbial species, commercially available as probiotics in ruminants. Their effects and modes of action on the targeted species are described later in this chapter.

3.3 Regulation

In the EU, Regulation EC 1831/2003 on feed additives covers microbial FA applied to animal nutrition, including ruminants. The regulation applies to all FA and premixtures, but not to processing aids or to veterinary medicinal products as defined by Directive 2001/82/EC. Only authorized additives may be placed on the market and used. Authorizations are granted for use in feed intended for specific animal species or categories and for specific conditions of use. Microbial FA are classified into the category of zootechnical additives, either as digestibility enhancers or gut flora stabilizers. Authorizations are valid for 10 years throughout the European Economic Area and they are renewable for 10-year periods. The authorization procedure is a long and strict process (Fig. 2) which has led to a very small number of microbial additives actually authorized in EU for ruminants (dairy cattle, beef cattle, small ruminants and growing ruminants).

Other countries possess their own regulation process which can also be complex and demanding. In the United States, the designation 'Generally Recognized As Safe' (GRAS) is authorized by the Food and Drug Administration (FDA) when a chemical or substance added to food is considered safe by experts, and is thus exempted from the usual Federal Food, Drug, and Cosmetic Act (FFDCA) food additive tolerance requirements. Marketed probiotics are considered GRAS. In other countries where there is no specific regulation, microbial FA are generally considered as safe but no specific demonstration of efficacy is required.

There is a growing concern about antimicrobial resistance (AMR) that might be developed by gut bacteria when exposed repeatedly to antibiotics. In this context, probiotic bacteria carrying AMR genes could be a problem for

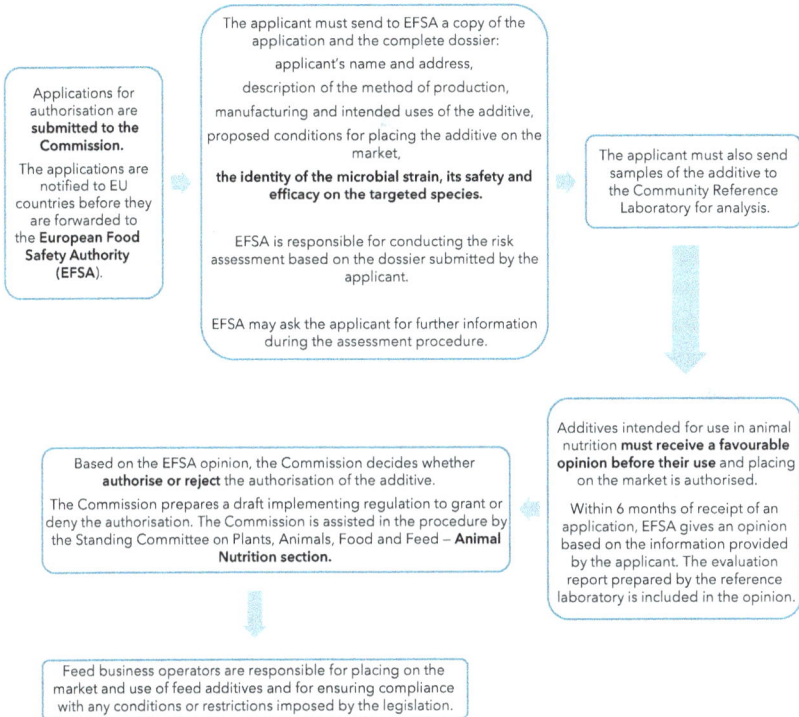

Figure 2 Current procedure for a microbial feed additive authorization for animal nutrition in EU (from Regulation (EC) N° 1831/2003 on additives for use in animal nutrition).

animal hosts. Amachawadi et al. (2018) have looked at commercial probiotics containing *Enterococcus faecium* for cattle and swine, in particular their sensitivity or resistance to antimicrobials and presence of virulence genes. They found that none of them harbored virulence genes. However, some exhibited AMR to medically important antimicrobials and were multi-resistant to chloramphenicol, erythromycin, penicillin, kanamycin, lincomycin and tetracycline, which suggests they might be a source of AMR in the animal gut. It is important to improve characterization of microbial strains used in ruminant feed, particularly by obtaining genomic information using whole genome sequencing (WGS), to ensure the complete safety of these microbials.

4 Benefits and modes of action of probiotics: young ruminants

As previously discussed, pre-weaning is the most critical period in the ruminant life cycle with the highest incidence of mortality and morbidity. Optimization of the pre-weaning phase is crucial to ensure successful herd management.

Microbial FA represent a promising opportunity for the dairy industry, particularly in the context of reducing antibiotic usage.

Currently available published trials do not report a consistent beneficial effect of microbial FA on performance (Alugongo et al., 2017; Fomenky et al., 2017; Galvão et al., 2005; He et al., 2017). This is probably due to variations in the design of studies, type of diet, FA strains and dosage, duration and mode of administration (in milk, in starter feed, or both, and combined or not with other additives), but also large individual variations observed among calves from the same herd. When microbial FA is given through starter feed, the low consumption of this solid feed generally observed during the first two weeks after birth prevents the calf from sufficient inoculation with live microbial cells, thus impacting efficacy of the supplement. For this reason, the distribution of microbial FA through milk has attracted as it allows a better control of the dose administered. When distributed in milk, the live probiotic reaches the intestine and can exert beneficial effects on gut health and the calf immune system.

As it has been noted, microbial inoculation of the rumen starts immediately at birth (Yeoman et al., 2018), and rapid changes occur in the composition of the ruminal bacterial community during the first days of life (Jami et al., 2013; Rey et al., 2012). In an experiment with lambs reared with their dams, the daily distribution of a live yeast additive (*Saccharomyces cerevisiae* CNCM I-1077) accelerated microbial establishment of functional communities in the rumen in very young lambs (Chaucheyras-Durand and Fonty, 2002). In gnotobiotically reared-lambs maintained in sterile isolators, the additive also stimulated the establishment of fibrolytic bacteria (Chaucheyras-Durand and Fonty, 2001). The same strain was found to improve microbial colonization in the maturing rumen of newborn artificially fed lambs, in particular ciliate protozoa, fungi and the fibrolytic bacterial species *Fibrobacter succinogenes* (Chaucheyras-Durand et al., unpublished data), with a potentially more specialized ecosystem toward efficient fiber degradation, as assessed by a dedicated DNA microarray (Comtet-Marre et al., 2018). This suggested a possible positive impact on fiber digestive efficiency (Chaucheyras-Durand, unpublished data). This yeast additive (*S. cerevisiae* CNCM I-1077) also had a beneficial effect in a calf performance trial up to 3 weeks post weaning, with growth and final weight both significantly increased, together with a better weight gain to feed ratio, probably because of improved microbial rumen maturation (Terré et al., 2015). Possible modes of action in the rumen may include an improved anaerobic environment, as redox potential of rumen contents of young lambs was shown to be lower with yeast supplementation, due to the capacity of live yeast cells to scavenge oxygen (Chaucheyras-Durand and Fonty, 2002). It may be also possible that yeast FA also supplies key nutrients, B vitamins or cofactors, able to promote growth and feed particle colonization by fibrolytic microbial populations (Chaucheyras-Durand et al., 2016).

It is noticeable that beneficial effects may depend on the initial health status of the young animal. A very good health status combined with optimal farm cleanliness and good husbandry, are among factors which may help to explain why no benefit of a microbial FA could be observed in the absence of any source of challenge or disease causing agents to the animals (Alugongo et al., 2017). In growth-retarded calves, the supplementation of *Bacillus* probiotics had a beneficial effect on body weight, feed intake, feed conversion and growth factors levels in serum. The volume of bacteria involved in production of energy and short chain fatty acids (such as *Proteobacteria*, *Rhodospirillaceae*, *Campylobacterales* and *Butyricimonas*) were increased, whereas undesirable mycoplasmas were decreased, compared to the non-supplemented group (Du et al., 2018). Administration of *Bacillus amyloliquefaciens* H57 to dairy calves was shown to increase weight gain although no significant differences were observed in rumen community structure. The low abundance of the *B. amyloliquefaciens* H57 in the rumen suggests that the probiotic was not directly responsible for weight gain but rather influenced animal behavior (feed consumption) or altered rumen community functions (Schofield et al., 2018).

Galvão et al. (2005) studied calves with low serum IgG concentrations due to lack of colostrum. Calves fed with milk replacer containing a *Saccharomyces boulardii* strain CNCM I-1079 had significantly less diarrhea and associated veterinary costs. *E. coli* is one of the main pathogens responsible for calf diarrhea and DFM are commonly used to alleviate diarrhea even in absence of the pathogen being identified. In a meta-analysis, Signorini et al. (2012) concluded that administration of a lactic acid bacteria (LAB) probiotic to pre-weaning dairy calves exerted a protective effect and reduced the incidence of diarrhea. Other studies have also shown prevention of diarrhea in dairy calves fed with *E. coli* Nissle 1917 (von Buenau et al., 2005) or with a mix of microorganisms (Mokhber-Dezfouli et al., 2007). A multispecies DFM bolus administrated to Holstein dairy calves suffering from diarrhea reduced duration of diarrhea (Renaud et al., 2019). A *Bacillus*-based DFM was evaluated in combination with an electrolyte given orally as a therapy against diarrhea in pre-weaned calves. Compared with the electrolyte alone, the associated additives did reduce *Clostridium perfringens* fecal shedding, decreased diarrhea severity and reduced treatment costs (Wehnes et al., 2009).

Recent findings have highlighted the potential positive role of the *S. boulardii* strain CNCM I-1079, distributed in milk replacer and in starter feed, on gut development in young dairy calves (Fomenky et al., 2017). Altered colon morphology and increased neutral mucin production were observed, suggesting earlier gut maturation in calves receiving the supplement. Acute phase proteins such as haptoglobin and C-reactive protein were also increased in the serum of probiotic-fed calves at weaning. Phagocytic activity of polymorphonuclear neutrophils isolated from plasma was also stimulated. These data suggest a possible role of *S. boulardii* CNCM I-1079 in enhancing the innate immune and inflammatory response of calves during the stressful weaning period.

In the same study, two probiotics (*S. boulardii* CNCM I-1079 and *L. acidophilus BT-1386*) were found to have a significant impact on gut bacteria community structure in the pre-weaning period, particularly in the ileum (Fomenky et al., 2018). Both probiotics significantly reduced the relative abundance of potential pathogenic bacteria *Streptococcus* and *Tyzzerella* and increased that of the beneficial fiber-degrading bacteria *Fibrobacter*. The live yeast supplementation seemed to have greater effects on gut microbiota than the bacterial DFM. DFM may impact bile acid secretion, known to affect fat digestion and absorption, as well as protein and energy metabolism and gut microbiota regulation. As shown in pigs, activation of bile acid-regulated pathways strengthens intestinal protection against bacterial infection and associated secretion of fluids and electrolytes, reduces inflammation in the colon and increases levels of the growth hormone FGF19 (Ipharraguerre et al., 2018). Different types of mechanisms of action for *S. boulardii* have been reported, though most have only been demonstrated *in vitro* or with laboratory animal experiments (Pothoulakis, 2009; McFarland, 2010; Stier and Bischoff, 2016). *S. boulardii* has been classified as having three modes of action: luminal, trophic and mucosal-anti-inflammatory signaling effects. Within the intestinal lumen, *S. boulardii* may interfere with pathogenic adhesion through yeast cell wall manno-oligosaccharides-bacterial fimbriae interaction. They could neutralize toxins, preserve cellular physiology, interact with intestinal microbiota and induce changes in the short chain fatty acid profile.

5 Benefits and modes of action of probiotics: feed efficiency in adult ruminants

Improving feed efficiency (FE) is important for sustainable livestock production. Recent metagenomics studies have shown associations between rumen microbiota composition/functions and FE (Mizrahi and Jami, 2018). There are particular associations between several plant-cell wall degrading bacteria taxa, their fiber-degrading function and FE (Delgado et al., 2019). Using a 16S rDNA sequencing approach, McGovern et al. (2018) showed a negative correlation between *Fibrobacter succinogenes* abundance and residual feed intake (RFI), suggesting this cellulolytic and hemicellulolytic bacteria contributes to FE by providing a substrate to the host and to other microbial populations. Improving fiber digestibility in the rumen is therefore a target for probiotics. It may be achieved by:

- reducing the indigestible fiber fraction, and
- increasing the rate of fiber digestion, for example, by maintaining a ruminal environment able to promote the population of fiber-digesting bacteria.

In forage indigestible fiber (iNDF) is generally related to lignin content and, in particular, structural carbohydrates (cellulose and hemicellulose) 'trapped' within lignin. Lignin is not fully digested in the animal gastrointestinal tract as its biochemical degradation process involves oxidative pathways. However, the release of lignin-bounded carbohydrates would potentially increase the feed value of forage. McSweeney et al. (1994) observed that up to 33.6% of sorghum lignin degradation could be associated with the activity of the ruminal anaerobic fungus *Neocallimastix patriciarum*. The authors suggested that the degraded lignin fraction was a lignin carbohydrate complex solubilized through dissolution of xylan from the matrix rather than through lignin depolymerization.

Microbial FA could improve fiber degradation in the rumen in different ways. An indirect route is through pH stabilization effects (discussed in the further section). Another route is modification of the ruminal environment through strengthening anaerobic conditions and oxygen-scavenging properties (Chaucheyras-Durand et al., 2008; Chaucheyras-Durand and Fonty, 2002; Jouany, 2006; Jouany and Morgavi, 2007; Marden et al., 2008). These conditions promote fiber-degrading microbiota and their effect on plant-cell wall polysaccharides. Specific nutritional requirements for vitamins, peptides, amino acids, ammonia, organic acids or branched-chain fatty acids have been described for bacteria and fungi, and it is likely that microbial FA could supply these components to fibrolytic microorganisms.

Research has identified the potential of live yeast FA to enhance growth and activity of fiber-degrading rumen microorganisms (Chaucheyras-Durand et al., 2008). Mechanisms include an increase in fungal zoospore germination and cellulose degradation, promotion of growth and/or activities of fibrolytic bacterial strains of *Fibrobacter succinogenes, Ruminococcus albus, Ruminococcus flavefaciens* and *Butyrivibrio fibrisolvens*. Several studies report increases in abundance of *F. succinogenes* (AlZahal et al., 2017; Pinloche et al., 2013; Uyeno et al., 2017), *Ruminococcus* (Mosoni et al., 2007; Pinloche et al., 2013; Silberberg et al., 2013; Sousa et al., 2018) or rumen fungi (Ding et al., 2014) using DNA-based techniques (qPCR, DNA sequencing). Jiang et al. (2017) compared the effect of live and heat-killed yeast FA on rumen microbiota using Illumina MiSeq and qPCR. The supplementation of live yeast increased the relative abundance of *Ruminococcus* and *F. succinogenes*. The impact of a live yeast strain (*S. cerevisiae* CNCM I-1077) was confirmed in promoting colonization of fibrous substrates by cellulolytic bacteria (*F. succinogenes, R. flavefaciens, B. fibrisolvens*) and fungi. It was observed that the degree of stimulation was dependent upon the nature of the substrate, and on the microbial species targeted (Chaucheyras-Durand et al., 2016). Feedstuffs with the highest levels of lignin and thereby with less easily accessible digestible carbohydrates were better degraded during yeast supplementation, suggesting an impact on the microbial breakdown of lignin-polysaccharide linkages, on which rumen fungi could be mostly active.

Live yeast has been found to promote the abundance of *Butyrivibrio fibrisolvens* on fibrous substrates. This bacterial species is known to possess ferulic and p-coumaric acid esterases which hydrolyze ester linkages between phenolic acids and xylan chains within the hemicellulose fraction, thereby exposing more polysaccharides to microbial enzymatic attack (McSweeney et al., 1998). Guedes et al. (2008) reported that the same live yeast strain increased fiber (NDF-neutral detergent fiber) degradation of corn silage samples. The yeast FA increased NDF degradation of the low-digestible corn silages more strongly than that of the high-digestible corn silage. These results suggest that live yeast could help to reduce indigestible NDF by promoting the action of bacteria and fungi involved in the hydrolysis of lignin-polysaccharide bonds.

Live yeast additives indirectly promote fiber degradation by stabilizing rumen pH in the case of ruminal acidosis. Using 18S and ITS sequencing, Ishaq et al. (2017) have shown that diet-induced subacute ruminal acidosis (SARA) modified the diversity of rumen fungi and protozoa and selected against fiber-degrading species. Cows supplemented with the live yeast product had a stabilized rumen pH and greater microbial diversity shifts which prevented a reduction in protozoa. The live yeast (*S. cerevisiae* Y1242) also increased the abundance of some dominant anaerobic OTU belonging to *F. succinogenes* and the abundance of genes encoding for specific microbial fibrolytic enzymes (AlZahal et al., 2017). It has been consistently reported that live yeast supplementation improves rumen fiber digestion *in vivo* (Chaucheyras-Durand et al., 2016; Dias et al., 2018; Ding et al., 2014; Ferraretto et al., 2012; Guedes et al., 2008; Sousa et al., 2018);

Ruminal acidosis is still commonly found in dairy and beef cattle, where a high amount of readily fermentable carbohydrates associated with low fiber in the diet may negatively impact rumen function due to high acid production and reduced buffering capacity (Villot et al., 2018). Although acidosis is not just a 'rumen disorder' but affects the total digestive tract (Plaizier et al., 2018), intervention strategies have focused on the rumen microbiota balance and rumen pH stabilization (Humer et al., 2018). There have been a number of studies of yeast FA in SARA prevention (Chaucheyras-Durand et al., 2016; Ipharraguerre et al., 2018; Jiang et al., 2017; Jouany, 2006; Jouany and Morgavi, 2007; Mizrahi and Jami, 2018). Rumen sensors (Villot et al., 2018) have been used to measure the beneficial effect of live yeast FA on ruminal pH.

Yeast FA promote shifts in microbial populations involved in release of fermentation acids (in particular lactate, which is a stronger acid than VFAs), and/or those implicated in lactic acid removal, leading to an optimized balance between lactate producers and lactate utilizers. Stimulation of growth and metabolism of lactate-utilizing bacteria, such as *Megasphaera elsdenii* or *Selenomonas ruminantium,* have been observed both *in vitro* in the presence of different live yeasts (Chaucheyras-Durand et al., 2008), and *in vivo* (Pinloche et al., 2013). The supply of growth factors, peptides, amino acids or vitamins has been proposed as a mechanism of action (Fonty and Chaucheyras-Durand, 2006). The inhibition of growth of *Streptococcus bovis*, one of the main bacterial species involved in lactate production, has been measured *in vitro* (Fonty and Chaucheyras-Durand, 2006). The impact of yeast FA on ruminal lactate concentration has been confirmed in *in vivo* studies (Chaucheyras-Durand et al., 2008; Kumprechtová et al., 2019; Reis et al., 2018).

Yeast FA can also alleviate butyric orientated acidosis (Brossard et al., 2004; Lettat et al., 2010). Brossard et al. (2006) reported the pH stabilizing effect of one strain of *S. cerevisiae* in sheep with butyric ruminal acidosis. This strain promoted ciliate Entodiniomorphid protozoa, which are known to engulf starch granules very rapidly and thus compete effectively with amylolytic bacteria (Owens et al., 1998). The effect of live yeast on ciliate protozoa has been reported in other studies (Chaucheyras-Durand and Fonty, 2002; Silberberg et al., 2013). The fact that ciliates digest starch at a slower rate than by amylolytic bacteria, and their main end-products of fermentation are VFAs rather than lactate, might explain why they have a stabilizing effect in the rumen by delaying fermentation. The promotion of ciliate protozoa, despite their association with methane production, could increase fiber digestibility as a recent rumen metatranscriptomic study showed that their contribution to fibrolysis appeared to be greater than previously thought (Comtet-Marre et al., 2017).

Better fiber digestion and stabilized rumen pH have been seen to benefit animal rumen health and its function by improvement of FE. De Ondarza et al. (2010) has investigated the effect of live yeast (*Saccharomyces cerevisiae* CNCM I-1077) on dairy cows by gathering performance data from 14 trials, clearly demonstrating that FE was improved. When targeting the cows fed diet above 30% NDF (high fiber diet, low SARA risk), FE was higher than the overall mean. The live yeast-treated animals produced an extra 40 g of milk per kg dry matter intake (DMI). In the case of a higher risk of SARA (low fiber diet, >25% starch), FE was even greater with an extra 80 g of milk per kg DMI. Other research has shown that when cows were fed with this live yeast product, eating behavior was modified with shorter intervals between meals (Bach et al., 2007; DeVries and Chevaux, 2014), indicating improvement in diet digestibility since intake was not affected. Improvement of rumen pH supports higher activity of fiber-degrading populations and thus explains the higher meal frequency.

A recent study using an endoscope to collect ruminal biopsies has evaluated the effect of a live yeast (*S. cerevisiae* CNCM I-1077) during the calving period (Bach et al., 2018). Lactate-producing *Streptococcus* and *Lactobacillus* genera (Derakhshani et al., 2016) and saccharolytic members of the Proteobacteria phylum (Zhu et al., 2017) increase post-partum, increasing the risk of SARA which affects the epithelium of the gastrointestinal tract (Steele et al., 2011). Results showed that supplementation of live yeast FA before calving increased expression of genes regulating inflammation and the epithelial barrier in the rumen (such as tight junction coding genes).

Few studies have looked at the potential of probiotics to alleviate inflammation or to control the immune system during acidosis. In heifers fed with high challenging diets (starch/fructose), a combination of the ionophore monensin and a live yeast FA (*S. cerevisiae* I-1077) was significantly reduced histamine concentration in rumen fluid (Golder et al., 2014). In other trials looking at the effect of microbial FA on rumen or plasma concentrations of inflammatory molecules, Garcia Diaz et al. (2018) have shown a decrease in plasmatic LPS concentrations as well as Serum Amyloid A (SAA), an acute phase protein, after a supplementation of steers by live yeasts (*S. cerevisiae* NCYC 996), but with no effect on ruminal or duodenal LPS concentrations. Silberberg et al. (2013) measured a significant decrease in plasma SAA levels with the use of live yeast FA, again with no effect on ruminal LPS concentrations. This may be because the large intestine rather than the rumen is the most probable site for LPS translocation in ruminants, because the monolayer intestinal epithelium is more prone to damage by acidity compared to the reticulo-rumen epithelium (Khiaosa-Ard and Zebeli, 2018).

Figure 3 summarizes the expected benefits of live yeast FA supplementation to ruminants at risk of subacute rumen acidosis (SARA).

When cattle are subjected to high temperature and humidity conditions (heat stress), the relative ruminal proportions of the *Clostridium*

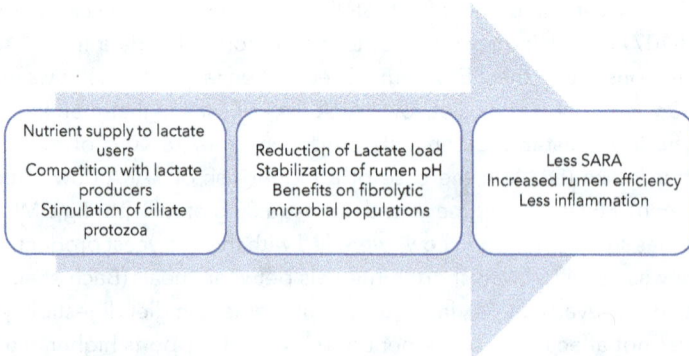

| Nutrient supply to lactate users Competition with lactate producers Stimulation of ciliate protozoa | Reduction of Lactate load Stabilization of rumen pH Benefits on fibrolytic microbial populations | Less SARA Increased rumen efficiency Less inflammation |

Figure 3 The benefits of live yeast FA in ruminants.

coccoides-*Eubacterium rectale* group and *Streptococcus* increase, while the genus *Fibrobacter* decrease, leading to microbial imbalance (Uyeno et al., 2010). Under these conditions, microbial FA could help to stabilize the rumen ecosystem and alleviate the negative impact of heat stress on cattle performance (Salvati et al., 2015).

Megasphaera elsdenii is as an ecologically important species within the rumen ecosystem as it removes lactate and thereby prevents acidosis. A recent metagenomic study, focusing on the rumen microbiome in low or high efficient cows, highlighted that genes that were the most enriched in the efficient cows' microbiomes were affiliated to *Megasphaera elsdenii*. Genes of the acrylate metabolic pathway, involved in the conversion of lactate into propionate were also enriched in the most efficient animals (Shabat et al., 2016). *M. elsdenii* has been considered as a FA to boost *M. elsdenii* concentrations in the rumen and speed up removal of lactate from the rumen (Arik et al., 2019; Muya et al., 2015; Yohe et al., 2018; Zebeli et al., 2012). However, it has not always been possible to measure beneficial effects (Yohe et al., 2018; Zebeli et al., 2012). In addition, preparation and delivery is challenging.

Other lactate-utilizing bacteria from the *Propionibacterium* genus have also been evaluated to alleviate the severity of SARA in high-grain fed cattle (Azad et al., 2017; Philippeau et al., 2017). It has been suggested associating them with *Lactobacilli* in order to promote lactic acid production, which would stimulate lactate-utilizing *Propionibacterium* (Lettat et al., 2012). A study using 16S rRNA gene sequencing reported beneficial effects of one *P. acidipropionici* strain P169 on the relative abundance of sequences affiliated to key lactate utilizers (*Veillonellaceae* and *Megasphaera),* and cellulolytic members of the bacterial families *Ruminococcaceae, Lachnospiraceae, Clostridiaceae, Christensenellaceae* which were enriched in the rumen microbiota of supplemented high-grain fed steers (Azad et al., 2017). Microbial metabolites were also affected (higher molar proportions of branched-chain fatty acids and increased concentration of ammonia) indicating an improved state of fibrolytic and proteolytic activity.

6 Benefits and modes of action of probiotics: methane production

The use of microbial FA is one possible option to decrease CH_4 emission from ruminants. Potential mechanisms of action include (Jeyanathan et al., 2014):

- direct inhibition of methanogenesis,
- promotion of alternative pathways which already exist in the rumen such as homoacetogenesis,
- fumarate reduction,
- propionate production through the acrylate pathway,
- nitrate/nitrite reduction,

- capnophily (CO_2 fixation), and
- anaerobic oxidation of methane (methanotrophy).

Only a few of these routes have been explored, probably because of a lack of understanding of these complex microbial metabolisms, and of their capacity to compete with methanogenesis. The homoacetogenic pathway has been found to promote functional ruminal activity in lambs deprived of methanogens but that acetogens were much less efficient in capturing H_2 than methanogens (Fonty et al., 2007). There are also problems of survival for candidate microbes since most of these alternative routes require strict anaerobic conditions. Finally, there are potential side-effects on fermentation efficiency or toxicity for the animal, for example, in the case of nitrite accumulation from nitrate supplementation which could cause methemoglobinaemia.

There have been several studies on the effect of *Saccharomyces cerevisiae* on rumen methanogenesis to promote alternative non-methanogenic pathways. The promotion of acetogenic bacteria has been shown at least *in vitro* (Chaucheyras et al., 1995; Nollet et al., 1997). Using gnotobiotic animal models, Chaucheyras-Durand et al. (2010b) showed that the composition of the cellulolytic community (hydrogen producers vs. non-hydrogen producers) may have an impact on H_2 accumulation and subsequent methane production in the rumen ecosystem. The promotion of fibrolytic organisms which do not produce any hydrogen, such as *Fibrobacter succinogenes*, may help limit methane emissions in the rumen. However, studies about yeast FA so far have failed to demonstrate a mitigating effect (Bayat et al., 2015; Chung et al., 2011). The increase in FE reported in the presence of live yeast FA (see earlier) should have an indirect effect on methane excretion, as it would decrease the amount of output/kg of milk/meat produced (Jeyanathan et al., 2014). A few bacteria have been tested for their anti-methanogenic potential. Results in sheep showed some efficacy with one *L. pentosus* strain (a 13% decrease in methane emission after 2 weeks which lasted throughout the 4 week-treatment) (Jeyanathan et al., 2016). A mix of three strains of *Propionibacterium* was also evaluated in beef cattle that fed a corn-finishing diet (Vyas et al., 2014) but showed no effect on enteric methane production; the authors suggested that the high starch content of the diet-induced high levels of propionate, and that these conditions may have reduced the *Propionibacterium* efficacy on methane mitigation.

7 Benefits and modes of action of probiotics: pathogen control

The use of antibiotics in livestock production has been linked to the development of resistant bacterial populations and the persistence of antibiotic residues in animal food products (Langford et al., 2003; Ramatla et al., 2017;

Seymour et al., 1988). Microbial FA supplementation has been considered as an alternative way to decrease pathogen loads in cattle as well as reduce the risk of transmission of zoonotic pathogens to humans. Salmonellosis was the second most important human zoonosis observed in the EU in 2017 (EFSA and ECDC, 2018), and *Salmonella* food-borne outbreaks have been associated with the consumption of beef products (EFSA and ECDC, 2018). *Salmonella* have been shown to be asymptomatically carried by cattle (Feye et al., 2016) although they are also capable of causing clinical diseases in animals. Diarrhea and enteric diseases caused by *Salmonella enterica* and *Escherichia coli* are a major cause of economic loss for cattle producers (USDA, 2014a; Cho and Yoon, 2014). Probiotics may help to eradicate pathogens through competitive exclusion, production of antimicrobial compounds and stimulation of the host immune defenses which will decrease pathogen colonization in the animal and the risk of infection.

Lactobacillus amylovorus C94 and *L. salivarius* C86 strains have demonstrated promising potential *in vitro* against *Salmonella enterica* isolated from cattle (Adetoye et al., 2018). DFM have been shown to have a positive effect in preventing *Salmonella* infection in steers treated daily with a mixture of *L. acidophilus* LA51 and *P. freudenreichii* PF24 (Tabe et al., 2008). Administration of the same DFM to beef cattle led to a significantly reduced prevalence of *Salmonella* at slaughter and a significant reduction of pathogen concentration in peripheral lymph nodes (PLN) associated with contamination of ground beef (Vipham et al., 2015). Soto et al. (2015) tested the effect of an inoculum composed of *Lactobacillus casei* DSPV318T, *L. salivarius* DSPV315T and *Pediococcus acidilactici* DSPV006T on young calves, and observed that daily administration of this inoculum coupled with lactose caused a decrease in the neutrophils/lymphocytes ratio, indicating an increase in immune response during the acute phase of infection (Soto et al., 2016). This product was previously observed to decrease severity of diarrhea in *Salmonella*-infected calves (Soto et al., 2015).

Shiga-toxin producing *E. coli* (STEC) (such as *E. coli* O157:H7) are of major concern due to their impact on human health (Caprioli et al., 2005; Chauret, 2011; EFSA and ECDC, 2018). DFM represent an efficient strategy to reduce fecal shedding of *E. coli* O157:H7 in beef cattle (Brashears and Chaves, 2017). *In vitro* it was shown that *Lactobacillus acidophilus* BT-1386 had a dose-dependent inhibitory effect on *E. coli* O157:H7 in feces (Chaucheyras-Durand et al., 2006). This strain and a live yeast strain of *Saccharomyces cerevisiae* CNCM I-1077 were able to significantly reduce pathogenic *E. coli* load in the rumen of sheep (Chaucheyras-Durand et al., 2010a). Suppression of *E. coli* O157:H7 was achieved in the rumen with *L. reuteri* LB1-7 coupled with glycerol. Reduced pathogen growth in rectal content was achieved following prior exposure to the same mixture in rumen fluid (Bertin et al., 2017). Those *in vitro* results highlight the potential of DFM to decrease pathogen load in the digestive system.

A recent meta-analysis has shown that LAB (particularly a DFM combination of *Lactobacillus acidophilus* NP51 and *Propionibacterium freudenreichii* NP24) reduces the prevalence of *E. coli* O157:H7 fecal shedding (Wisener et al., 2015). In 2013, The Beef Industry Food Safety Council (BIFSCO) in the United States has recognized the efficiency of several bacterial strains in reducing *E. coli* O157:H7 in beef cattle and include them as part of Production Best Practice (Beef Industry Food Safety Council subcommittee on pre-harvest, 2013). Administration of a mixture of competitive-exclusion commensal *E. coli* to experimentally infected calves showed a significant reduction of fecal shedding for O157:H7 but also non-O157 serotypes of *E. coli* (Tkalcic et al., 2003). Using the same probiotic mixture, another study found treated calves shed significantly less non-O157 *E. coli* but no effect was observed on *E. coli* O157 (Zhao et al., 2003). In experimentally infected lambs, daily administration of *S. faecium* or a mixture of *S. faecium*, *L. acidophilus*, *L. casei*, *L. fermentum* and *L. plantarum* led to significant reductions in *E. coli* O157:H7 in feces (Lema et al., 2001). A probiotic mixture containing *Lactobacillus acidophilus* LC10, *Lactobacillus helveticus* LC3, *Lactobacillus bulgaricus* LC182, *Lactobacillus lactis*, *Streptococcus thermophilus* LC201 and *Enterococcus faecium* LAT E-253 was shown to significantly reduce the fecal shedding of pathogenic *E. coli* in sheep (Rigobelo et al., 2015). However, the limitation of STEC carriage is dependent upon many parameters which are difficult to control (such as diet, environmental factors, stress level, intermittent and seasonal shedding, occurrence of several serotypes within the herd, and so on (Bertin et al., 2011, 2013; Chaucheyras-Durand et al., 2010a, 2006; Dunière et al., 2011; Fremaux et al., 2006). Identification and implementation of the most efficient strategies should be based on a good understanding of STEC ecology and physiology in bovine GIT (Segura et al., 2018).

Probiotics can also be administered to cattle to treat non-digestive disorders. One of the most detrimental diseases in dairy industry is mastitis as it leads to decrease in production and higher treatment costs (Hogeveen and Østerås, 2005). Probiotics such as LAB can potentially reduce mastitis through colonizing the udder and building a beneficial biofilm that prevents pathogen colonization. Three strains of *Lactobacillus brevis* 1595 and 1597 and *Lactobacillus plantarum* 1610 have shown good colonization capacities in competing with pathogens in mammary gland colonization. They also exhibit anti-inflammatory properties, with lower IL-8 secretion by *E. coli*-stimulated bovine mammary epithelial cells (bMEC) (Bouchard et al., 2015). *Lactobacillus rhamnosus* ATCC7469 and *L. plantarum* 2/37 have been shown to form biofilms to displace *Staphylococcus* (Wallis et al., 2018, 2019). A bacteriocin-producing *Lactococcus lactis* DPC3174 was shown to be as effective as a conventional antibiotic to treat cow mastitis

(Klostermann et al., 2008). Two LAB strains (*Lactobacillus perolens* CRL 1724 and nisin Z producer *Lactococcus lactis* sub. *Lactis* CRL1655) were shown to modulate the host udder immune system and stimulate local and systemic defence lines *in vivo* in dairy cow at the dry-off stage (Pellegrino et al., 2017). In their review of the recent literature on the use of probiotics to treat bovine mastitis, Rainard and Foucras (2018) point that most trials with LAB are *in vitro* rather than *in vivo*. However, they confirm intra-mammary administration as the best approach for mastitis control with probiotics.

Probiotics can also be used in order to enhance reproduction performance. Bovine reproductive diseases can lead to lower milk production as well as impaired reproductive performance (Bellows et al., 2002). Acute metritis is an inflammation of the uterus after calving due to bacterial infection, with *E. coli* being the predominant pathogen initiating the disease (Kassé et al., 2016). Vaginal treatment with LAB probiotic was observed to reduce metritis prevalence up to 58% compared to control animals (Genís et al., 2018). Live yeast dietary supplementation has been shown to improve reproductive performance of dairy cows during heat stress through the alteration of hormones and ovarian follicular dynamics (Nasiri et al., 2018).

Bovine respiratory disease (BRD) results in significant losses for cattle producers (Fulton et al., 2002; Griffin et al., 2010). BRD is multifactorial but one of the predominant bacterial pathogens identified is *Mannheimia haemolytica*. Probiotic strains belonging to *Lactobacillus*, *Lactococcus* and *Paenibacillus* genera have been shown to adhere to bovine respiratory cells and inhibit *M. haemolytica* through competition and displacement *in vitro* (Amat et al., 2017). Intragastric administration of *Enterococcus faecalis* CECT7121 to mice was shown to induce an increase in humoral immune response against *Pasteurella multocida* and *M. haemolytica*. Probiotic-treated mice showed higher interferon-γ production indicating a higher cellular immune response (Díaz et al., 2018). The authors suggested that this probiotic could be used as a possible adjuvant in a vaccine strategy to enhance ruminant immune response. In beef heifers submitted to a bacterial and viral challenge (*M. haemolytica* and bovine herpesvirus-1), the supplementation of *Saccharomyces boulardii* CNCM I-1079 altered the leukogram (with a significant increase in neutrophil %, and an increase in monocyte % on the day following *Mannheimia* challenge), indicating an increase in the innate immune response (Kayser et al., 2019).

8 Benefits and modes of action of probiotics: effects on the immune system

As seen earlier in this chapter, there is significant interest in probiotic intervention to enhance immune function in neonate and pre-weaned ruminants. Research demonstrates that gut microbiota in early life has a key role in the shaping of immune-competence later in life (Gensollen et al., 2016). In adult ruminants, probiotics may play a role in modulating the production of pro- vs. anti-inflammatory signals in immune function (Raabis et al., 2019). However, the way probiotics participate in host-microbial interaction and modulate immune function remains largely unknown (Ma et al., 2018). Modulation of the immune response associated with supplementation of the diet by probiotics may occur not only through the innate and adaptive immune system, but also through regulation of intestinal epithelium permeability, mucus secretion and competition within the microbial ecosystem where probiotics can secrete antimicrobial compounds (La Fata et al., 2018). This makes the picture very complex (Fig. 4).

Bach et al. (2018) assessed the effect of a live yeast FA (*S. cerevisiae* I-1077) on the expression of genes encoding for proteins involved in the immune response in cows around calving. Cows responded rapidly to the live yeast, with measurable changes in expression of genes involved in the immune response after 7 days of supplementation, including the expression of the anti-inflammatory IL-10 genes in the rumen and of the β-defensin gene in the colon. β-glucans or manno-oligosaccharides (e.g. *Saccharomyces cerevisiae*) have also been considered as immunostimulants or pathogen binders (Ganner and Schatzmayr, 2012; Goodridge et al., 2009; Li et al., 2016, 2018; Yuan et al., 2015). Other species of yeast such as *Debaryomyces hansenii* are also of growing interest as the cell wall β-glucans seem to induce a marked immune response in other animal species (Angulo et al., 2018, 2019). There has been great interest in improving the understanding of the role of yeast cell wall components on immune system modulation and host response using techniques such as atomic force microscopy (Schiavone et al., 2015, 2017).

There is a significant body of literature on the effect of selected bacteria on intestinal epithelial cell (IEC) response, mostly from human cell lines or in rodent models (La Fata et al., 2018). Villena et al. (2018) have suggested the use of bovine IEC to study the impact of immunobiotics, such as *Bifidobacteria* or *Lactobacillus*, on anti-viral immunity. They have reported beneficial effects of immunobiotics through activation of interferon regulatory factor-3 (IRF-3), increased production of anti-viral factors and regulation of inflammation mediators. The impact of bacterial probiotics (*Lactobacillus casei* BL23, *Lactococcus lactis* V7) has been studied on the innate immune response of bovine mammary cells stimulated by *Staphylococcus aureus* (Assis et al., 2015; Souza et al., 2018). Like yeasts bacteria cell wall composition and organization are likely to be involved in triggering host response (Lebeer et al., 2018).

There is a growing interest in the use of neurochemical-producing probiotics in the treatment of health and disease through the microbiota-gut-brain axis (Lyte, 2011). This is significant for ruminants since there is increasing

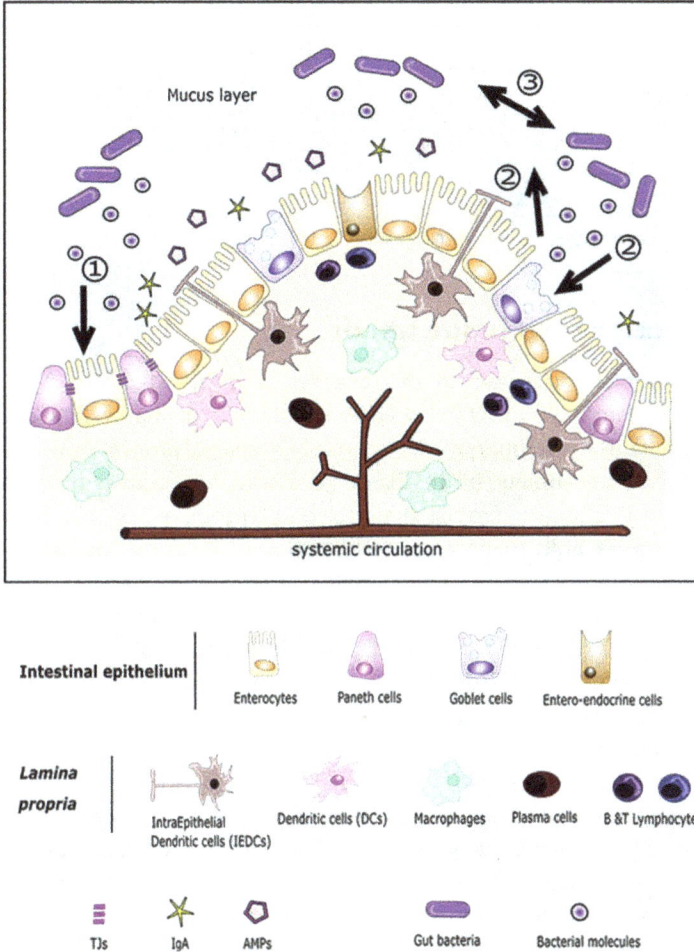

Figure 4 The intestinal barrier and the main actors involved in microbiota (gut bacteria or probiotics and their metabolites) – intestinal epithelium cross-talk. ① represents modulation of tight junction (TJ) proteins, ② shows the modulation of mucus secretion from goblet cells, ③ shows possible interactions between different members of gut microbiota (antimicrobial properties for example). Source: adapted from La Fata et al. (2018).

evidence to show that stress increases susceptibility to enteric infections (Freestone and Lyte, 2010). Recent findings report production of dopamine by *Enterococcus faecium* strains acting in the gastrointestinal tract (Villageliú and Lyte, 2018). Other microbial species used as probiotics are also able to produce a range of molecules involved in gut-brain axis communication (Lyte, 2011) (Table 2).

Table 2 Neurochemicals isolated by potential probiotic genera

Microbial genus	Neurochemical
Bifidobacterium, Lactobacillus	Gamma amino butyric acid (GABA)
Lactobacillus	Acetylcholine
Bacillus, Saccharomyces	Norepinephrine (Noradrenaline)
Enterococcus, Bacillus	Dopamine
Enterococcus	Serotonine

Source: adapted from Lyte (2011) and Villageliú and Lyte (2018).

9 Conclusions and future trends

As the research discussed in this chapter suggests, microbial FA use in livestock can improve profitability while maintaining animal welfare and farm sustainability. The evolution of commercially available probiotics will depend on regulatory constraints. In the EU, for example, the focus is on approving a single, well-defined microbial strain for one target species. It would help if regulators were able to accept combinations of microbial actives or even uncharacterized microbial consortia if they had proven benefits. Innovative techniques, such as microbial transplants (ruminal or fecal), are currently not allowed, nor are genetically modified probiotics authorized.

More research is needed to better understand the mechanisms by which microbial FA interact with gut microbiota and the host animal, in order to better select safe, robust, environmentally friendly and efficient additives which can be used easily by farmers. Multi-omic approaches will aid in deciphering metabolic pathways utilized by probiotics which can explain their benefits in the host animal. Once these pathways are characterized, it will be easier to use this understanding for more accurate and appropriate screening and selection of the best candidate for a given expected benefit. Most ruminant research has been focused on improving rumen function for FE or methane mitigation. However, there needs to be more research on how probiotics modulate host-microbial interactions and host immunity in order to optimize probiotic intervention strategies to improve animal gut health and productivity.

Other potential areas of research include maternal imprinting (on GIT microbiota establishment and immunity development in offspring) through microbial FA, although mechanisms of action remain largely unknown and require further research. Another exciting field is microbial endocrinology, a discipline at the interface of microbiology, endocrinology and neurophysiology. This can help understand better how stress affects ruminants health. To quote Freestone: 'happy, less stressed ruminants may be better-nourished animals and safer sources of meat' (Freestone and Lyte, 2010).

10 Acknowledgements

We are very grateful to Ms Raphaële Gresse (Lallemand/UMR MEDIS) for her skilled contribution in the preparation of Fig. 4.

11 Where to look for further information section

Further readings to get deeper information on the topics which were covered in this chapter:

- Chaucheyras-Durand, F. and Durand, H. 2010. Probiotics in animal nutrition and health. *Benef Microbes*. 1(1), 3–9. doi:10.3920/BM2008.1002 (Review).
- McCann, J. C., Elolimy, A. A. and Loor, J. J. 2017. Rumen microbiome, probiotics, and fermentation additives. *Vet. Clin. North Am. Food. Anim. Pract*. 33(3), 539–53. doi:10.1016/j.cvfa.2017.06.009.
- McSweeney, C. and Mackie, R. 2012. Micro-organisms and ruminant digestion: state of knowledge, trends and future prospects. Commission on Genetic Resources for Food and Agriculture. Background study paper NO. 61.
- Probiotics in animal nutrition. FAO report 2016. Available at: http://www.fao.org/3/a-i5933e.pdf.

Key societies, professional organisations or other websites worth visiting to keep up to date with trends:

- www.lallemandanimalnutrition.com.
- www.ruminantdigestivesystem.com.
- www6.ara.inrae.fr/medis.

Key journals or conferences (e.g. key conferences held regularly every year or few years):

- Beneficial Microbes Conference (https://www.bastiaanse-communicati on.com/BMC2018/) and associated journal.
- Congress on Gastro-intestinal Function (https://www.congressgastrofun ction.org/).
- International Probiotic Conference (https://www.probiotic-conference. net/).
- Joint INRA-Rowett Symposium on gut microbiota (https://colloque.inrae. fr/inra-rowett-2016).

12 References

Abaker, J. A., Xu, T. L., Jin, D., Chang, G. J., Zhang, K. and Shen, X. Z. 2017. Lipopolysaccharide derived from the digestive tract provokes oxidative stress in the liver of dairy cows fed a high-grain diet. *J. Dairy Sci.* 100(1), 666–78. doi:10.3168/jds.2016-10871.

Abecia, L., Martín-García, A. I., Martínez, G., Newbold, C. J. and Yáñez-Ruiz, D. R. 2013. Nutritional intervention in early life to manipulate rumen microbial colonization and methane output by kid goats postweaning. *J. Anim. Sci.* 91(10), 4832–40. doi:10.2527/jas.2012-6142.

Abecia, L., Martínez-Fernandez, G., Waddams, K., Martín-García, A. I., Pinloche, E., Creevey, C. J., Denman, S. E., Newbold, C. J. and Yáñez-Ruiz, D. R. 2014. An antimethanogenic nutritional intervention in early life of ruminants modifies ruminal colonization by Archaea. *Archaea* 2014, 841463. doi:10.1155/2014/841463.

Abecia, L., Martínez-Fernandez, G., Waddams, K., Martín-García, A. I., Pinloche, E., Creevey, C. J., Denman, S. E., Newbold, C. J. and Yáñez-Ruiz, D. R. 2018. Analysis of the rumen microbiome and metabolome to study the effect of an antimethanogenic treatment applied in early life of kid goats. *Front. Microbiol.* 9, 2227. doi:10.3389/fmicb.2018.02227.

Adetoye, A., Pinloche, E., Adeniyi, B. A. and Ayeni, F. A. 2018. Characterization and anti-salmonella activities of lactic acid bacteria isolated from cattle faeces. *BMC Microbiol.* 18(1), 96. doi:10.1186/s12866-018-1248-y.

Alugongo, G. M., Xiao, J., Wu, Z., Li, S., Wang, Y. and Cao, Z. 2017. Review: utilization of yeast of Saccharomyces cerevisiae origin in artificially raised calves. *J. Anim. Sci. Biotechnol.* 8, 34. doi:10.1186/s40104-017-0165-5.

AlZahal, O., Li, F., Guan, L. L., Walker, N. D. and McBride, B. W. 2017. Factors influencing ruminal bacterial community diversity and composition and microbial fibrolytic enzyme abundance in lactating dairy cows with a focus on the role of active dry yeast. *J. Dairy Sci.* 100(6), 4377–93. doi:10.3168/jds.2016-11473.

Amachawadi, R. G., Giok, F., Shi, X., Soto, J., Narayanan, S. K., Tokach, M. D., Apley, M. D. and Nagaraja, T. G. 2018. Antimicrobial resistance of *Enterococcus faecium* strains isolated from commercial probiotic products used in cattle and swine. *J. Anim. Sci.* 96(3), 912–20. doi:10.1093/jas/sky056.

Amat, S., Subramanian, S., Timsit, E. and Alexander, T. W. 2017. Probiotic bacteria inhibit the bovine respiratory pathogen *Mannheimia haemolytica* serotype 1 *in vitro*. *Lett. Appl. Microbiol.* 64(5), 343–9. doi:10.1111/lam.12723.

Angulo, M., Reyes-Becerril, M., Tovar-Ramírez, D., Ascencio, F. and Angulo, C. 2018. *Debaryomyces hansenii* CBS 8339 β-glucan enhances immune responses and down-stream gene signaling pathways in goat peripheral blood leukocytes. *Dev. Comp. Immunol.* 88, 173–82. doi:10.1016/j.dci.2018.07.017.

Angulo, M., Reyes-Becerril, M., Cepeda-Palacios, R., Tovar-Ramírez, D., Esteban, M. Á. and Angulo, C. 2019. Probiotic effects of marine *Debaryomyces hansenii* CBS 8339 on innate immune and antioxidant parameters in newborn goats. *Appl. Microbiol. Biotechnol.* 103(5), 2339–52 doi:10.1007/s00253-019-09621-5.

Arik, H. D., Gulsen, N., Hayirli, A. and Alatas, M. S. 2019. Efficacy of Megasphaera elsdenii inoculation in subacute ruminal acidosis in cattle. *J. Anim. Physiol. Anim. Nutr.* 103(2), 416–26 doi:10.1111/jpn.13034.

Ashenafi, D., Yidersal, E., Hussen, E., Solomon, T. and Desiye, M. 2018. The effect of long distance transportation stress on cattle: a review. *Biomed. J. Sci. Tech. Res.* 3. doi:10.26717/BJSTR.2018.03.000908.

Assis, B. S., Germon, P., Silva, A. M., Even, S., Nicoli, J. R. and Le Loir, Y. 2015. *Lactococcus lactis* V7 inhibits the cell invasion of bovine mammary epithelial cells by *Escherichia coli* and *Staphylococcus aureus*. *Benef Microbes* 6(6), 879–86. doi:10.3920/BM2015.0019.

Azad, E., Narvaez, N., Derakhshani, H., Allazeh, A. Y., Wang, Y., McAllister, T. A. and Khafipour, E. 2017. Effect of *Propionibacterium acidipropionici* P169 on the rumen and faecal microbiota of beef cattle fed a maize-based finishing diet. *Benef. Microbes* 8(5), 785–99. doi:10.3920/BM2016.0145.

Bach, A., Iglesias, C. and Devant, M. 2007. Daily rumen pH pattern of loose-housed dairy cattle as affected by feeding pattern and live yeast supplementation. *Anim. Feed Sci. Technol.* 136(1–2), 146–53. doi:10.1016/j.anifeedsci.2006.09.011.

Bach, A., Guasch, I., Elcoso, G., Chaucheyras-Durand, F., Castex, M., Fàbregas, F., Garcia-Fruitos, E. and Aris, A. 2018. Changes in gene expression in the rumen and colon epithelia during the dry period through lactation of dairy cows and effects of live yeast supplementation. *J. Dairy Sci.* 101(3), 2631–40. doi:10.3168/jds.2017-13212.

Baldwin, R. L., McLeod, K. R., Klotz, J. L. and Heitmann, R. N. 2004. Rumen development, intestinal growth and hepatic metabolism in the pre- and postweaning ruminant. *J. Dairy Sci.* 87, E55–65. doi:10.3168/jds.S0022-0302(04)70061-2.

Bayat, A. R., Kairenius, P., Stefański, T., Leskinen, H., Comtet-Marre, S., Forano, E., Chaucheyras-Durand, F. and Shingfield, K. J. 2015. Effect of Camelina oil or live yeasts (*Saccharomyces cerevisiae*) on ruminal methane production, rumen fermentation, and milk fatty acid composition in lactating cows fed grass silage diets. *J. Dairy Sci.* 98(5), 3166–81. doi:10.3168/jds.2014-7976.

Beef Industry Food Safety Council subcommittee on pre-harvest. 2013. Production Best Practices (PBP) to aid in the control of foodborne pathogens in groups of cattle. BIFSCO, USA.

Bellows, D. S., Ott, S. L. and Bellows, R. A. 2002. Review: Cost of reproductive diseases and conditions in cattle. *Prof. Anim. Sci.* 18(1), 26–32. doi:10.15232/S1080-7446(15)31480-7.

Bertin, Y., Girardeau, J. P., Chaucheyras-Durand, F., Lyan, B., Pujos-Guillot, E., Harel, J. and Martin, C. 2011. Enterohaemorrhagic *Escherichia coli* gains a competitive advantage by using ethanolamine as a nitrogen source in the bovine intestinal content. *Environ. Microbiol.* 13(2), 365–77. doi:10.1111/j.1462-2920.2010.02334.x.

Bertin, Y., Chaucheyras-Durand, F., Robbe-Masselot, C., Durand, A., de la Foye, A., Harel, J., Cohen, P. S., Conway, T., Forano, E. and Martin, C. 2013. Carbohydrate utilization by enterohaemorrhagic *Escherichia coli* O157:H7 in bovine intestinal content. *Environ. Microbiol.* 15(2), 610–22. doi:10.1111/1462-2920.12019.

Bertin, Y., Habouzit, C., Dunière, L., Laurier, M., Durand, A., Duchez, D., Segura, A., Thévenot-Sergentet, D., Baruzzi, F., Chaucheyras-Durand, F. and Forano, E. 2017. *Lactobacillus reuteri* suppresses *E. coli* O157:H7 in bovine ruminal fluid: toward a pre-slaughter strategy to improve food safety? *PLoS ONE* 12(11), e0187229. doi:10.1371/journal.pone.0187229.

Blecha, F., Boyles, S. L. and Riley, J. G. 1984. Shipping suppresses lymphocyte blastogenic responses in Angus and Brahman × Angus feeder calves. *J. Anim. Sci.* 59(3), 576–83. doi:10.2527/jas1984.593576x.

Bouchard, D. S., Seridan, B., Saraoui, T., Rault, L., Germon, P., Gonzalez-Moreno, C., Nader-Macias, F. M., Baud, D., François, P., Chuat, V., Chain, F., Langella, P., Nicoli, J., Le Loir, Y. and Even, S. 2015. Lactic acid bacteria isolated from bovine mammary microbiota: potential allies against bovine mastitis. *PLoS ONE* 10(12), e0144831. doi:10.1371/journal.pone.0144831.

Brashears, M. M. and Chaves, B. D. 2017. The diversity of beef safety: a global reason to strengthen our current systems. *Meat Sci.* 132, 59-71. doi:10.1016/j.meatsci.2017.03.015.

Brossard, L., Martin, C., Chaucheyras-Durand, F. and Michalet-Doreau, B. 2004. Protozoa involved in butyric rather than lactic fermentative pattern during latent acidosis in sheep. *Reprod. Nutr. Dev.* 44(3), 195-206. doi:10.1051/rnd:2004023.

Brossard, L., Chaucheyras-Durand, F., Michalet-Doreau, B. and Martin, C. 2006. Dose effect of live yeasts on rumen microbial communities and fermentations during butyric latent acidosis in sheep: new type of interaction. *Anim. Sci.* 82(6), 829-36. doi:10.1017/ASC200693.

Bush, S. J., McCulloch, M. E. B., Muriuki, C., Salavati, M., Davis, G. M., Farquhar, I. L., Lisowski, Z. M., Archibald, A. L., Hume, D. A. and Clark, E. L. 2019. Comprehensive transcriptional profiling of the gastrointestinal tract of ruminants from birth to adulthood reveals strong developmental stage specific gene expression. *G3 Bethesda MD* 9(2), 359-73. doi:10.1534/g3.118.200810.

Caprioli, A., Morabito, S., Brugère, H. and Oswald, E. 2005. Enterohaemorrhagic *Escherichia coli*: emerging issues on virulence and modes of transmission. *Vet. Res.* 36(3), 289-311. doi:10.1051/vetres:2005002.

Chaucheyras, F., Fonty, G., Bertin, G. and Gouet, P. 1995. *In vitro* H2 utilization by a ruminal acetogenic bacterium cultivated alone or in association with an archaea methanogen is stimulated by a probiotic strain of Saccharomyces cerevisiae. *Appl. Environ. Microbiol.* 61(9), 3466-7.

Chaucheyras-Durand, F. and Fonty, G. 2001. Establishment of cellulolytic bacteria and development of fermentative activities in the rumen of gnotobiotically-reared lambs receiving the microbial additive *Saccharomyces cerevisiae* CNCM I-1077. *Reprod. Nutr. Dev.* 41(1), 57-68. doi:10.1051/rnd:2001112.

Chaucheyras-Durand, F. and Fonty, G. 2002. Influence of a probiotic yeast (*Saccharomyces cerevisiae* CNCM I-1077) on microbial colonization and fermentations in the rumen of newborn lambs. *Microb. Ecol. Health Dis.* 14(1), 30-6. doi:10.1080/089106002760002739.

Chaucheyras-Durand, F., Madic, J., Doudin, F. and Martin, C. 2006. Biotic and abiotic factors influencing *in vitro* growth of *Escherichia coli* O157:H7 in ruminant digestive contents. *Appl. Environ. Microbiol.* 72(6), 4136-42. doi:10.1128/AEM.02600-05.

Chaucheyras-Durand, F., Walker, N. D. and Bach, A. 2008. Effects of active dry yeasts on the rumen microbial ecosystem: past, present and future. *Anim. Feed Sci. Technol.* 145(1-4), 5-26. doi:10.1016/j.anifeedsci.2007.04.019.

Chaucheyras-Durand, F., Faqir, F., Ameilbonne, A., Rozand, C. and Martin, C. 2010a. Fates of acid-resistant and non-acid-resistant Shiga toxin-producing *Escherichia coli* strains in ruminant digestive contents in the absence and presence of probiotics. *Appl. Environ. Microbiol.* 76(3), 640-7. doi:10.1128/AEM.02054-09.

Chaucheyras-Durand, F., Masséglia, S., Fonty, G. and Forano, E. 2010b. Influence of the composition of the cellulolytic flora on the development of hydrogenotrophic microorganisms, hydrogen utilization, and methane production in the rumens of gnotobiotically reared lambs. *Appl. Environ. Microbiol.* 76(24), 7931-7. doi:10.1128/AEM.01784-10.

Chaucheyras-Durand, F., Ameilbonne, A., Bichat, A., Mosoni, P., Ossa, F. and Forano, E. 2016. Live yeasts enhance fibre degradation in the cow rumen through an increase in plant substrate colonization by fibrolytic bacteria and fungi. *J. Appl. Microbiol.* 120(3), 560-70. doi:10.1111/jam.13005.

Chauret, C. 2011. Survival and control of *Escherichia coli* O157:H7 in foods, beverages, soil and water. *Virulence* 2(6), 593-601. doi:10.4161/viru.2.6.18423.

Chen, Y., Penner, G. B., Li, M., Oba, M. and Guan, L. L. 2011. Changes in bacterial diversity associated with epithelial tissue in the beef cow rumen during the transition to a high-grain diet. *Appl. Environ. Microbiol.* 77(16), 5770-81. doi:10.1128/AEM.00375-11.

Chen, Y., Oba, M. and Guan, L. L. 2012. Variation of bacterial communities and expression of toll-like receptor genes in the rumen of steers differing in susceptibility to subacute ruminal acidosis. *Vet. Microbiol.* 159(3-4), 451-9. doi:10.1016/j.vetmic.2012.04.032.

Cho, Y. I. and Yoon, K. J. 2014. An overview of calf diarrhea - infectious etiology, diagnosis, and intervention. *J. Vet. Sci.* 15(1), 1-17. doi:10.4142/jvs.2014.15.1.1.

Chung, Y. H., Walker, N. D., McGinn, S. M. and Beauchemin, K. A. 2011. Differing effects of 2 active dried yeast (*Saccharomyces cerevisiae*) strains on ruminal acidosis and methane production in nonlactating dairy cows. *J. Dairy Sci.* 94(5), 2431-9. doi:10.3168/jds.2010-3277.

Comtet-Marre, S., Parisot, N., Lepercq, P., Chaucheyras-Durand, F., Mosoni, P., Peyretaillade, E., Bayat, A. R., Shingfield, K. J., Peyret, P. and Forano, E. 2017. Metatranscriptomics reveals the active bacterial and eukaryotic fibrolytic communities in the rumen of dairy cow fed a mixed diet. *Front. Microbiol.* 8, 67. doi:10.3389/fmicb.2017.00067.

Comtet-Marre, S., Chaucheyras-Durand, F., Bouzid, O., Mosoni, P., Bayat, A. R., Peyret, P. and Forano, E. 2018. FibroChip, a functional DNA microarray to monitor cellulolytic and hemicellulolytic activities of rumen microbiota. *Front. Microbiol.* 9, 215. doi:10.3389/fmicb.2018.00215.

Delgado, B., Bach, A., Guasch, I., González, C., Elcoso, G., Pryce, J. E. and Gonzalez-Recio, O. 2019. Whole rumen metagenome sequencing allows classifying and predicting feed efficiency and intake levels in cattle. *Sci. Rep.* 9(1), 11. doi:10.1038/s41598-018-36673-w.

de Ondarza, M. B., Sniffen, C. J., Dussert, L., Chevaux, E., Sullivan, J. and Walker, N. 2010. Case Study: multiple-study analysis of the effect of live yeast on milk yield, milk component content and yield, and feed efficiency. *Prof. Anim. Sci.* 26(6), 661-6. doi:10.15232/S1080-7446(15)30664-1.

Derakhshani, H., Tun, H. M., Cardoso, F. C., Plaizier, J. C., Khafipour, E. and Loor, J. J. 2016. Linking peripartal dynamics of ruminal microbiota to dietary changes and production parameters. *Front. Microbiol.* 7, 2143. doi:10.3389/fmicb.2016.02143.

DeVries, T. J. and Chevaux, E. 2014. Modification of the feeding behavior of dairy cows through live yeast supplementation. *J. Dairy Sci.* 97(10), 6499-510. doi:10.3168/jds.2014-8226.

Dias, A. L. G., Freitas, J. A., Micai, B., Azevedo, R. A., Greco, L. F. and Santos, J. E. P. 2018. Effect of supplemental yeast culture and dietary starch content on rumen fermentation and digestion in dairy cows. *J. Dairy Sci.* 101(1), 201-21. doi:10.3168/jds.2017-13241.

Díaz, A. M., Almozni, B., Molina, M. A., Sparo, M. D., Manghi, M. A., Canellada, A. M. and Castro, M. S. 2018. Potentiation of the humoral immune response elicited by a commercial vaccine against bovine respiratory disease by *Enterococcus faecalis* CECT7121. *Benef Microbes* 9(4), 553–62. doi:10.3920/BM2017.0081.

Ding, G., Chang, Y., Zhao, L., Zhou, Z., Ren, L. and Meng, Q. 2014. Effect of *Saccharomyces cerevisiae* on alfalfa nutrient degradation characteristics and rumen microbial populations of steers fed diets with different concentrate-to-forage ratios. *J. Anim. Sci. Biotechnol.* 5(1), 24. doi:10.1186/2049-1891-5-24.

Du, R., Jiao, S., Dai, Y., An, J., Lv, J., Yan, X., Wang, J. and Han, B. 2018. Probiotic *Bacillus amyloliquefaciens* C-1 improves growth performance, stimulates GH/IGF-1, and regulates the gut microbiota of growth-retarded beef calves. *Front. Microbiol.* 9, 2006. doi:10.3389/fmicb.2018.02006.

Dunière, L., Gleizal, A., Chaucheyras-Durand, F., Chevallier, I. and Thévenot-Sergentet, D. 2011. Fate of *Escherichia coli* O26 in corn silage experimentally contaminated at ensiling, at silo opening, or after aerobic exposure, and protective effect of various bacterial inoculants. *Appl. Environ. Microbiol.* 77(24), 8696–704. doi:10.1128/AEM.06320-11.

Durand-Chaucheyras, F., Fonty, G., Bertin, G., Théveniot, M. and Gouet, P. 1998. Fate of Levucell SC I-1077 yeast additive during digestive transit in lambs. *Reprod. Nutr. Dev.* 38(3), 275–80. doi:10.1051/rnd:19980307.

EFSA, ECDC. 2018. The European Union summary report on trends and sources of zoonoses, zoonotic agents and food-borne outbreaks in 2017 [WWW Document]. European Food Safety Authority. Available at: https://www.efsa.europa.eu/fr/efsajournal/pub/5500 (accessed on 03 June 2019).

Ferraretto, L. F., Shaver, R. D. and Bertics, S. J. 2012. Effect of dietary supplementation with live-cell yeast at two dosages on lactation performance, ruminal fermentation, and total-tract nutrient digestibility in dairy cows. *J. Dairy Sci.* 95(7), 4017–28. doi:10.3168/jds.2011-5190.

Feye, K. M., Anderson, K. L., Scott, M. F., Henry, D. L. and Dorton, K. L. 2016. Abrogation of *Salmonella* and *E. coli* O157:H7 in feedlot cattle fed a proprietary *Saccharomyces cerevisiae* fermentation prototype. *J. Vet. Sci. Technol.* 7(4) doi:10.4172/2157-7579.1000350.

Fomenky, B. E., Chiquette, J., Bissonnette, N., Talbot, G., Chouinard, P. Y. and Ibeagha-Awemu, E. M. 2017. Impact of *Saccharomyces cerevisiae* boulardii CNCMI-1079 and *Lactobacillus acidophilus* BT1386 on total lactobacilli population in the gastrointestinal tract and colon histomorphology of Holstein dairy calves. *Anim. Feed Sci. Technol.* 234, 151–61. doi:10.1016/j.anifeedsci.2017.08.019.

Fomenky, B. E., Do, D. N., Talbot, G., Chiquette, J., Bissonnette, N., Chouinard, Y. P., Lessard, M. and Ibeagha-Awemu, E. M. 2018. Direct-fed microbial supplementation influences the bacteria community composition of the gastrointestinal tract of pre- and post-weaned calves. *Sci. Rep.* 8(1), 14147. doi:10.1038/s41598-018-32375-5.

Fonty, G. and Chaucheyras-Durand, F. 2006. Effects and modes of action of live yeasts in the rumen. *Biol. (Bratisl.)* 61(6), 741–50. doi:10.2478/s11756-006-0151-4.

Fonty, G., Joblin, K., Chavarot, M., Roux, R., Naylor, G. and Michallon, F. 2007. Establishment and development of ruminal hydrogenotrophs in methanogen-free lambs. *Appl. Environ. Microbiol.* 73(20), 6391–403. doi:10.1128/AEM.00181-07.

Freestone, P. and Lyte, M. 2010. Stress and microbial endocrinology: prospects for ruminant nutrition. *Animal* 4(7), 1248–57. doi:10.1017/S1751731110000674.

Fremaux, B., Raynaud, S., Beutin, L. and Rozand, C. V. 2006. Dissemination and persistence of Shiga toxin-producing *Escherichia coli* (STEC) strains on French dairy farms. *Vet. Microbiol.* 117(2–4), 180–91. doi:10.1016/j.vetmic.2006.04.030.

Fulton, R. W., Cook, B. J., Step, D. L., Confer, A. W., Saliki, J. T., Payton, M. E., Burge, L. J., Welsh, R. D. and Blood, K. S. 2002. Evaluation of health status of calves and the impact on feedlot performance: assessment of a retained ownership program for postweaning calves. *Can. J. Vet. Res.* 66(3), 173–80.

Gaggìa, F., Mattarelli, P. and Biavati, B. 2010. Probiotics and prebiotics in animal feeding for safe food production. *Int. J. Food Microbiol.* 141(Suppl. 1), S15–28. doi:10.1016/j.ijfoodmicro.2010.02.031.

Galvão, K. N., Santos, J. E. P., Coscioni, A., Villaseñor, M., Sischo, W. M. and Berge, A. C. 2005. Effect of feeding live yeast products to calves with failure of passive transfer on performance and patterns of antibiotic resistance in fecal *Escherichia coli*. *Reprod. Nutr. Dev.* 45(4), 427–40. doi:10.1051/rnd:2005040.

Ganner, A. and Schatzmayr, G. 2012. Capability of yeast derivatives to adhere enteropathogenic bacteria and to modulate cells of the innate immune system. *Appl. Microbiol. Biotechnol.* 95(2), 289–97. doi:10.1007/s00253-012-4140-y.

Garcia Diaz, T., Ferriani Branco, A., Jacovaci, F. A., Cabreira Jobim, C., Bolson, D. C. and Pratti Daniel, J. L. 2018. Inclusion of live yeast and mannan-oligosaccharides in high grain-based diets for sheep: ruminal parameters, inflammatory response and rumen morphology. *PLoS ONE* 13(2), e0193313. doi:10.1371/journal.pone.0193313.

Genís, S., Cerri, R. L. A., Bach, À., Silper, B. F., Baylão, M., Denis-Robichaud, J. and Arís, A. 2018. Pre-calving intravaginal administration of lactic acid bacteria reduces metritis prevalence and regulates blood neutrophil gene expression after calving in dairy cattle. *Front. Vet. Sci.* 5, 135. doi:10.3389/fvets.2018.00135.

Gensollen, T., Iyer, S. S., Kasper, D. L. and Blumberg, R. S. 2016. How colonization by microbiota in early life shapes the immune system. *Science* 352(6285), 539–44. doi:10.1126/science.aad9378.

Golder, H. M., Celi, P., Rabiee, A. R. and Lean, I. J. 2014. Effects of feed additives on rumen and blood profiles during a starch and fructose challenge. *J. Dairy Sci.* 97(2), 985–1004. doi:10.3168/jds.2013-7166.

Goodridge, H. S., Wolf, A. J. and Underhill, D. M. 2009. Beta-glucan recognition by the innate immune system. *Immunol. Rev.* 230(1), 38–50. doi:10.1111/j.1600-065X.2009.00793.x.

Griffin, D., Chengappa, M. M., Kuszak, J. and McVey, D. S. 2010. Bacterial pathogens of the bovine respiratory disease complex. *Vet. Clin. North Am. Food Anim. Pract.* 26(2), 381–94. doi:10.1016/j.cvfa.2010.04.004.

Guedes, C. M., Gonçalves, D., Rodrigues, M. A. M. and Dias-da-Silva, A. 2008. Effects of a Saccharomyces cerevisiae yeast on ruminal fermentation and fibre degradation of maize silages in cows. *Anim. Feed Sci. Technol.* 145(1–4), 27–40. doi:10.1016/j.anifeedsci.2007.06.037.

Guzman, C. E., Bereza-Malcolm, L. T., De Groef, B. and Franks, A. E. 2015. Presence of selected methanogens, Fibrolytic bacteria, and proteobacteria in the gastrointestinal tract of neonatal dairy calves from birth to 72 hours. *PLoS ONE* 10(7), e0133048. doi:10.1371/journal.pone.0133048.

Guzman, C. E., Bereza-Malcolm, L. T., De Groef, B. and Franks, A. E. 2016. Uptake of milk with and without solid feed during the monogastric phase: effect on fibrolytic and methanogenic microorganisms in the gastrointestinal tract of calves. *Anim. Sci. J.* 87(3), 378–88. doi:10.1111/asj.12429.

Hare, K. S., Leal, L. N., Romao, J. M., Hooiveld, G. J., Soberon, F., Berends, H., Van Amburgh, M. E., Martín-Tereso, J. and Steele, M. A. 2019. Preweaning nutrient supply alters mammary gland transcriptome expression relating to morphology, lipid accumulation, DNA synthesis, and RNA expression in Holstein heifer calves. *J. Dairy Sci.* 102(3), 2618–30. doi:10.3168/jds.2018-15699.

He, Z. X., Ferlisi, B., Eckert, E., Brown, H. E., Aguilar, A. and Steele, M. A. 2017. Supplementing a yeast probiotic to pre-weaning Holstein calves: feed intake, growth and fecal biomarkers of gut health. *Anim. Feed Sci. Technol.* 226, 81–7. doi:10.1016/j.anifeedsci.2017.02.010.

Hogeveen, H. and Østerås, O. 2005. Mastitis management in an economic framework. In: *Mastitis in Dairy Production.* Wageningen Academic Publishers, Maastrich, Netherlands. doi:10.3920/978-90-8686-550-5.

Humer, E., Petri, R. M., Aschenbach, J. R., Bradford, B. J., Penner, G. B., Tafaj, M., Südekum, K. H. and Zebeli, Q. 2018. Invited review: practical feeding management recommendations to mitigate the risk of subacute ruminal acidosis in dairy cattle. *J. Dairy Sci.* 101(2), 872–88. doi:10.3168/jds.2017-13191.

Hutcheson, D. P. and Cole, N. A. 1986. Management of transit-stress syndrome in cattle: nutritional and environmental effects. *J. Anim. Sci.* 62(2), 555–60. doi:10.2527/jas1986.622555x.

Ingvartsen, K. L., Dewhurst, R. J. and Friggens, N. C. 2003. On the relationship between lactational performance and health: is it yield or metabolic imbalance that cause production diseases in dairy cattle? A position paper. *Livest. Prod. Sci.* 83(2–3), 277–308. doi:10.1016/S0301-6226(03)00110-6.

Ipharraguerre, I. R., Pastor, J. J., Gavaldà-Navarro, A., Villarroya, F. and Mereu, A. 2018. Antimicrobial promotion of pig growth is associated with tissue-specific remodeling of bile acid signature and signaling. *Sci. Rep.* 8(1), 13671. doi:10.1038/s41598-018-32107-9.

Ishaq, S. L., AlZahal, O., Walker, N. and McBride, B. 2017. An investigation into rumen fungal and protozoal diversity in three rumen fractions, during high-fiber or grain-induced sub-acute ruminal acidosis conditions, with or without active dry yeast supplementation. *Front. Microbiol.* 8, 1943. doi:10.3389/fmicb.2017.01943.

Jami, E., Israel, A., Kotser, A. and Mizrahi, I. 2013. Exploring the bovine rumen bacterial community from birth to adulthood. *ISME J.* 7(6), 1069–79. doi:10.1038/ismej.2013.2.

Jami, E., White, B. A. and Mizrahi, I. 2014. Potential role of the bovine rumen microbiome in modulating milk composition and feed efficiency. *PLoS ONE* 9(1), e85423. doi:10.1371/journal.pone.0085423.

Jeyanathan, J., Martin, C. and Morgavi, D. P. 2014. The use of direct-fed microbials for mitigation of ruminant methane emissions: a review. *Animal* 8(2), 250–61. doi:10.1017/S1751731113002085.

Jeyanathan, J., Martin, C. and Morgavi, D. P. 2016. Screening of bacterial direct-fed microbials for their antimethanogenic potential *in vitro* and assessment of their effect on ruminal fermentation and microbial profiles in sheep. *J. Anim. Sci.* 94(2), 739–50. doi:10.2527/jas.2015-9682.

Jiang, Y., Ogunade, I. M., Qi, S., Hackmann, T. J., Staples, C. R. and Adesogan, A. T. 2017. Effects of the dose and viability of Saccharomyces cerevisiae. 1. Diversity of ruminal microbes as analyzed by Illumina MiSeq sequencing and quantitative PCR. *J. Dairy Sci.* 100(1), 325–42. doi:10.3168/jds.2016-11263.

Jiao, J., Li, X., Beauchemin, K. A., Tan, Z., Tang, S. and Zhou, C. 2015. Rumen development process in goats as affected by supplemental feeding v. grazing: age-related anatomic development, functional achievement and microbial colonisation. *Br. J. Nutr.* 113(6), 888–900. doi:10.1017/S0007114514004413.

Johnsen, J. F., Viljugrein, H., Bøe, K. E., Gulliksen, S. M., Beaver, A., Grøndahl, A. M., Sivertsen, T. and Mejdell, C. M. 2019. A cross-sectional study of suckling calves' passive immunity and associations with management routines to ensure colostrum intake on organic dairy farms. *Acta Vet. Scand.* 61(1), 7. doi:10.1186/s13028-019-0442-8.

Jouany, J. P. 2006. Optimizing rumen functions in the close-up transition period and early lactation to drive dry matter intake and energy balance in cows. *Anim. Reprod. Sci.* 96(3–4), 250–64. doi:10.1016/j.anireprosci.2006.08.005.

Jouany, J. P. and Morgavi, D. P. 2007. Use of 'natural' products as alternatives to antibiotic feed additives in ruminant production. *Animal* 1(10), 1443–66. doi:10.1017/S1751731107000742.

Kassé, F. N., Fairbrother, J. M. and Dubuc, J. 2016. Relationship between *Escherichia coli* virulence factors and postpartum metritis in dairy cows. *J. Dairy Sci.* 99(6), 4656–67. doi:10.3168/jds.2015-10094.

Kayser, W. C., Carstens, G. E., Washburn, K. E., Welsh, T. H., Lawhon, S. D., Reddy, S. M., Pinchak, W. E., Chevaux, E. and Skidmore, A. L. 2019. Effects of combined viral-bacterial challenge with or without supplementation of *Saccharomyces cerevisiae* boulardii strain CNCM I-1079 on immune upregulation and DMI in beef heifers. *J. Anim. Sci.* 97(3), 1171–84 doi:10.1093/jas/sky483.

Khafipour, E., Krause, D. O. and Plaizier, J. C. 2009. A grain-based subacute ruminal acidosis challenge causes translocation of lipopolysaccharide and triggers inflammation. *J. Dairy Sci.* 92(3), 1060–70. doi:10.3168/jds.2008-1389.

Khan, M. A., Bach, A., Weary, D. M. and von Keyserlingk, M. A. G. 2016. Invited review: transitioning from milk to solid feed in dairy heifers. *J. Dairy Sci.* 99(2), 885–902. doi:10.3168/jds.2015-9975.

Khiaosa-Ard, R. and Zebeli, Q. 2018. Diet-induced inflammation: from gut to metabolic organs and the consequences for the health and longevity of ruminants. *Res. Vet. Sci.* 120, 17–27. doi:10.1016/j.rvsc.2018.08.005.

Kleen, J. L., Hooijer, G. A., Rehage, J. and Noordhuizen, J. P. T. M. 2003. Subacute ruminal acidosis (SARA): a review. *J. Vet. Med. A* 50(8), 406–14. doi:10.1046/j.1439-0442.2003.00569.x.

Klostermann, K., Crispie, F., Flynn, J., Ross, R. P., Hill, C. and Meaney, W. 2008. Intramammary infusion of a live culture of *Lactococcus lactis* for treatment of bovine mastitis: comparison with antibiotic treatment in field trials. *J. Dairy Res.* 75(3), 365–73. doi:10.1017/S0022029908003373.

Kumprechtová, D., Illek, J., Julien, C., Homolka, P., Jančík, F. and Auclair, E. 2019. Effect of live yeast (*Saccharomyces cerevisiae*) supplementation on rumen fermentation and metabolic profile of dairy cows in early lactation. *J. Anim. Physiol. Anim. Nutr.* 103(2), 447–55. doi:10.1111/jpn.13048.

La Fata, G., Weber, P. and Mohajeri, M. H. 2018. Probiotics and the gut immune system: indirect regulation. *Probiotics Antimicrob. Proteins* 10(1), 11–21. doi:10.1007/s12602-017-9322-6.

Langford, F. M., Weary, D. M. and Fisher, L. 2003. Antibiotic resistance in gut bacteria from dairy calves: a dose response to the level of antibiotics fed in milk. *J. Dairy Sci.* 86(12), 3963–6. doi:10.3168/jds.S0022-0302(03)74006-5.

Lebeer, S., Bron, P. A., Marco, M. L., Van Pijkeren, J.-P., O'Connell Motherway, M., Hill, C., Pot, B., Roos, S. and Klaenhammer, T. 2018. Identification of probiotic effector molecules: present state and future perspectives. *Curr. Opin. Biotechnol.* 49, 217–23. doi:10.1016/j.copbio.2017.10.007.

Lema, M., Williams, L. and Rao, D. R. 2001. Reduction of fecal shedding of enterohemorrhagic *Escherichia coli* O157:H7 in lambs by feeding microbial feed supplement. *Small Rumin. Res.* 39(1), 31–9. doi:10.1016/s0921-4488(00)00168-1.

Lettat, A., Nozière, P., Silberberg, M., Morgavi, D. P., Berger, C. and Martin, C. 2010. Experimental feed induction of ruminal lactic, propionic, or butyric acidosis in sheep. *J. Anim. Sci.* 88(9), 3041–46. doi:10.2527/jas.2010-2926.

Lettat, A., Nozière, P., Silberberg, M., Morgavi, D. P., Berger, C. and Martin, C. 2012. Rumen microbial and fermentation characteristics are affected differently by bacterial probiotic supplementation during induced lactic and subacute acidosis in sheep. *BMC Microbiol.* 12, 142. doi:10.1186/1471-2180-12-142.

Li, Z., You, Q., Ossa, F., Mead, P., Quinton, M. and Karrow, N. A. 2016. Assessment of yeast *Saccharomyces cerevisiae* component binding to *Mycobacterium avium* subspecies paratuberculosis using bovine epithelial cells. *BMC Vet. Res.* 12, 42. doi:10.1186/s12917-016-0665-0.

Li, Z., Kang, H., You, Q., Ossa, F., Mead, P., Quinton, M. and Karrow, N. A. 2018. *In vitro* bioassessment of the immunomodulatory activity of *Saccharomyces cerevisiae* components using bovine macrophages and *Mycobacterium avium* ssp. paratuberculosis. *J. Dairy Sci.* 101(7), 6271–86. doi:10.3168/jds.2017-13805.

Lyte, M. 2011. Probiotics function mechanistically as delivery vehicles for neuroactive compounds: microbial endocrinology in the design and use of probiotics. *Bioessays* 33(8), 574–81. doi:10.1002/bies.201100024.

Ma, T., Suzuki, Y. and Guan, L. L. 2018. Dissect the mode of action of probiotics in affecting host-microbial interactions and immunity in food producing animals. *Vet. Immunol. Immunopathol.* 205, 35–48. doi:10.1016/j.vetimm.2018.10.004.

Malmuthuge, N., Li, M., Fries, P., Griebel, P. J. and Guan, L. L. 2012. Regional and age dependent changes in gene expression of toll-like receptors and key antimicrobial defence molecules throughout the gastrointestinal tract of dairy calves. *Vet. Immunol. Immunopathol.* 146(1), 18–26. doi:10.1016/j.vetimm.2012.01.010.

Malmuthuge, N., Chen, Y., Liang, G., Goonewardene, L. A. and Guan, L. 2015. Heat-treated colostrum feeding promotes beneficial bacteria colonization in the small intestine of neonatal calves. *J. Dairy Sci.* 98(11), 8044–53. doi:10.3168/jds.2015-9607.

Marden, J. P., Julien, C., Monteils, V., Auclair, E., Moncoulon, R. and Bayourthe, C. 2008. How does live yeast differ from sodium bicarbonate to stabilize ruminal pH in high-yielding dairy cows? *J. Dairy Sci.* 91(9), 3528–35. doi:10.3168/jds.2007-0889.

McFarland, L. V. 2010. Systematic review and meta-analysis of *Saccharomyces boulardii* in adult patients. *World J. Gastroenterol.* 16(18), 2202–22. doi:10.3748/wjg.v16.i18.2202.

McGovern, E., Kenny, D. A., McCabe, M. S., Fitzsimons, C., McGee, M., Kelly, A. K. and Waters, S. M. 2018. 16S rRNA sequencing reveals relationship Between potent cellulolytic genera and feed efficiency in the rumen of bulls. *Front. Microbiol.* 9, 1842. doi:10.3389/fmicb.2018.01842.

McGuirk, S. M. 2008. Disease management of dairy calves and heifers. *Vet. Clin. North Am. Food Anim. Pract.* 24, 139–53. doi:10.1016/j.cvfa.2007.10.003.

McGuirk, S. M. and Collins, M. 2004. Managing the production, storage, and delivery of colostrum. *Vet. Clin. North Am. Food Anim. Pract.* 20, 593–603. doi:10.1016/j.cvfa.2004.06.005.

McSweeney, C. S., Dulieu, A., Katayama, Y. and Lowry, J. B. 1994. Solubilization of lignin by the ruminal anaerobic fungus *Neocallimastix patriciarum*. *Appl. Environ. Microbiol.* 60(8), 2985–89.

McSweeney, C. S., Dulieu, A. and Bunch, R. 1998. *Butyrivibrio* spp. and other xylanolytic microorganisms from the rumen have *Cinnamoyl esterase* activity. *Anaerobe* 4(1), 57–65. doi:10.1006/anae.1997.0130.

Meale, S. J., Li, S., Azevedo, P., Derakhshani, H., Plaizier, J. C., Khafipour, E. and Steele, M. A. 2016. Development of ruminal and fecal microbiomes are affected by weaning but not weaning strategy in dairy calves. *Front. Microbiol.* 7, 582. doi:10.3389/fmicb.2016.00582.

Meale, S. J., Chaucheyras-Durand, F., Berends, H., Guan, L. L. and Steele, M. A. 2017. From pre- to postweaning: transformation of the young calf's gastrointestinal tract. *J. Dairy Sci.* 100(7), 5984–95. doi:10.3168/jds.2016-12474.

Minuti, A., Palladino, A., Khan, M. J., Alqarni, S., Agrawal, A., Piccioli-Capelli, F., Hidalgo, F., Cardoso, F. C., Trevisi, E. and Loor, J. J. 2015. Abundance of ruminal bacteria, epithelial gene expression, and systemic biomarkers of metabolism and inflammation are altered during the peripartal period in dairy cows. *J. Dairy Sci.* 98(12), 8940–51. doi:10.3168/jds.2015-9722.

Mizrahi, I. and Jami, E. 2018. Review: The compositional variation of the rumen microbiome and its effect on host performance and methane emission. *Animal* 12(s2), s220–32. doi:10.1017/S1751731118001957.

Mokhber-Dezfouli, M. R., Tajik, P., Bolourchi, M. and Mahmoudzadeh, H. 2007. Effects of probiotics supplementation in daily milk intake of newborn calves on body weight gain, body height, diarrhea occurrence and health condition. *Pak. J. Biol. Sci.* 10(18), 3136–40. doi:10.3923/pjbs.2007.3136.3140.

Morelli, L. and Capurso, L. 2012. FAO/WHO guidelines on probiotics: 10 years later. *J. Clin. Gastroenterol.* 46(Suppl.), S1–2. doi:10.1097/MCG.0b013e318269fdd5.

Mosoni, P., Chaucheyras-Durand, F., Béra-Maillet, C. and Forano, E. 2007. Quantification by real-time PCR of cellulolytic bacteria in the rumen of sheep after supplementation of a forage diet with readily fermentable carbohydrates: effect of a yeast additive. *J. Appl. Microbiol.* 103(6), 2676–85. doi:10.1111/j.1365-2672.2007.03517.x.

Mõtus, K., Viltrop, A. and Emanuelson, U. 2018. Reasons and risk factors for beef calf and youngstock on-farm mortality in extensive cow-calf herds. *Animal* 12(9), 1958–66. doi:10.1017/S1751731117003548.

Muya, M. C., Nherera, F. V., Miller, K. A., Aperce, C. C., Moshidi, P. M. and Erasmus, L. J. 2015. Effect of *Megasphaera elsdenii* NCIMB 41125 dosing on rumen development, volatile fatty acid production and blood β-hydroxybutyrate in neonatal dairy calves. *J. Anim. Physiol. Anim. Nutr.* 99(5), 913–8. doi:10.1111/jpn.12306.

Nasiri, A. H., Towhidi, A., Shakeri, M., Zhandi, M., Dehghan-Banadaky, M. and Colazo, M. G. 2018. Effects of live yeast dietary supplementation on hormonal profile, ovarian follicular dynamics, and reproductive performance in dairy cows exposed to high ambient temperature. *Theriogenology* 122, 41–6. doi:10.1016/j.theriogenology.2018.08.013.

Nollet, L., Demeyer, D. and Verstraete, W. 1997. Effect of 2-bromoethanesulfonic acid and Peptostreptococcus productus ATCC 35244 addition on stimulation of reductive acetogenesis in the ruminal ecosystem by selective inhibition of methanogenesis. *Appl. Environ. Microbiol.* 63(1), 194–200.

Owens, F. N., Secrist, D. S., Hill, W. J. and Gill, D. R. 1998. Acidosis in cattle: a review. *J. Anim. Sci.* 76(1), 275–86. doi:10.2527/1998.761275x.

Pellegrino, M., Berardo, N., Giraudo, J., Nader-Macías, M. E. F. and Bogni, C. 2017. *Bovine mastitis* prevention: humoral and cellular response of dairy cows inoculated with lactic acid bacteria at the dry-off period. *Benef. Microbes* 8(4), 589–96. doi:10.3920/BM2016.0194.

Petri, R. M., Schwaiger, T., Penner, G. B., Beauchemin, K. A., Forster, R. J., McKinnon, J. J. and McAllister, T. A. 2013a. Changes in the rumen epimural bacterial diversity of beef cattle as affected by diet and induced ruminal acidosis. *Appl. Env. Microbiol.* 79, 3744–55. doi:10.1128/AEM.03983-12.

Petri, R. M., Schwaiger, T., Penner, G. B., Beauchemin, K. A., Forster, R. J., McKinnon, J. J. and McAllister, T. A. 2013b. Characterization of the core rumen microbiome in cattle during transition from forage to concentrate as well as during and after an acidotic challenge. *PLoS ONE* 8(12), e83424. doi:10.1371/journal.pone.0083424.

Petri, R. M., Kleefisch, M. T., Metzler-Zebeli, B. U., Zebeli, Q. and Klevenhusen, F. 2018. Changes in the rumen epithelial microbiota of cattle and host gene expression in response to alterations in dietary carbohydrate composition. *Appl. Environ. Microbiol.* 84(12). doi:10.1128/AEM.00384-18.

Philippeau, C., Lettat, A., Martin, C., Silberberg, M., Morgavi, D. P., Ferlay, A., Berger, C. and Nozière, P. 2017. Effects of bacterial direct-fed microbials on ruminal characteristics, methane emission, and milk fatty acid composition in cows fed high- or low-starch diets. *J. Dairy Sci.* 100(4), 2637–50. doi:10.3168/jds.2016-11663.

Pinloche, E., McEwan, N., Marden, J. P., Bayourthe, C., Auclair, E. and Newbold, C. J. 2013. The effects of a probiotic yeast on the bacterial diversity and population structure in the rumen of cattle. *PLoS ONE* 8(7), e67824. doi:10.1371/journal.pone.0067824.

Pitta, D. W., Indugu, N., Baker, L., Vecchiarelli, B. and Attwood, G. 2018. Symposium review: understanding diet-microbe interactions to enhance productivity of dairy cows. *J. Dairy Sci.* 101(8), 7661–79. doi:10.3168/jds.2017-13858.

Plaizier, J. C., Khafipour, E., Li, S., Gozho, G. N. and Krause, D. O. 2012. Subacute ruminal acidosis (SARA), endotoxins and health consequences. *Anim. Feed Sci. Technol.* 172, 9–21. doi:10.1016/j.anifeedsci.2011.12.004.

Plaizier, J. C., Danesh Mesgaran, M., Derakhshani, H., Golder, H., Khafipour, E., Kleen, J. L., Lean, I., Loor, J., Penner, G. and Zebeli, Q. 2018. Review: enhancing gastrointestinal health in dairy cows. *Animal* 12(s2), s399–418. doi:10.1017/S1751731118001921.

Pothoulakis, C. 2009. Review article: anti-inflammatory mechanisms of action of *Saccharomyces boulardii. Aliment. Pharmacol. Ther.* 30(8), 826–33, Article: Anti-Inflammatory. doi:10.1111/j.1365-2036.2009.04102.x.

Priestley, D., Bittar, J. H., Ibarbia, L., Risco, C. A. and Galvão, K. N. 2013. Effect of feeding maternal colostrum or plasma-derived or colostrum-derived colostrum replacer on passive transfer of immunity, health, and performance of preweaning heifer calves. *J. Dairy Sci.* 96(5), 3247–56. doi:10.3168/jds.2012-6339.

Raabis, S., Li, W. and Cersosimo, L. 2019. Effects and immune responses of probiotic treatment in ruminants. *Vet. Immunol. Immunopathol.* 208, 58–66. doi:10.1016/j.vetimm.2018.12.006.

Rainard, P. and Foucras, G. 2018. A critical appraisal of probiotics for mastitis control. *Front. Vet. Sci.* 5, 251. doi:10.3389/fvets.2018.00251.

Ramatla, T., Ngoma, L., Adetunji, M. and Mwanza, M. 2017. Evaluation of antibiotic residues in raw meat using different analytical methods. *Antibiot. Basel Switz* 6(4). doi:10.3390/antibiotics6040034.

Reis, L. F., Sousa, R. S., Oliveira, F. L. C., Rodrigues, F. A. M. L., Araújo, C. A. S. C., Meira-Júnior, E. B. S., Barrêto-Júnior, R. A., Mori, C. S., Minervino, A. H. H. and Ortolani, E. L. 2018. Comparative assessment of probiotics and monensin in the prophylaxis of acute ruminal lactic acidosis in sheep. *BMC Vet. Res.* 14(1), 9. doi:10.1186/s12917-017-1264-4.

Renaud, D. L., Kelton, D. F., Weese, J. S., Noble, C. and Duffield, T. F. 2019. Evaluation of a multispecies probiotic as a supportive treatment for diarrhea in dairy calves: a randomized clinical trial. *J. Dairy Sci.* 102(5), 4498–505 doi:10.3168/jds.2018-15793.

Rey, M., Enjalbert, F. and Monteils, V. 2012. Establishment of ruminal enzyme activities and fermentation capacity in dairy calves from birth through weaning. *J. Dairy Sci.* 95(3), 1500–12. doi:10.3168/jds.2011-4902.

Rey, M., Enjalbert, F., Combes, S., Cauquil, L., Bouchez, O. and Monteils, V. 2014. Establishment of ruminal bacterial community in dairy calves from birth to weaning is sequential. *J. Appl. Microbiol.* 116(2), 245–57. doi:10.1111/jam.12405.

Reynolds, J. D. and Morris, B. 1984. The effect of antigen on the development of Peyer's patches in sheep. *Eur. J. Immunol.* 14(1), 1–6. doi:10.1002/eji.1830140102.

Riaz, Q. U. A. and Masud, T. 2013. Recent trends and applications of encapsulating materials for probiotic stability. *Crit. Rev. Food Sci. Nutr.* 53(3), 231–44. doi:10.1080/10408398.2010.524953.

Rigobelo, E. E. C., Karapetkov, N., Maestá, S. A., Avila, F. A. and McIntosh, D. 2015. Use of probiotics to reduce faecal shedding of Shiga toxin-producing *Escherichia coli* in sheep. *Benef. Microbes* 6(1), 53–60. doi:10.3920/BM2013.0094.

Roth, B. A., Keil, N. M., Gygax, L. and Hillmann, E. 2009. Influence of weaning method on health status and rumen development in dairy calves. *J. Dairy Sci.* 92(2), 645–56. doi:10.3168/jds.2008-1153.

Russell, J. B. and Wilson, D. B. 1996. Why are ruminal cellulolytic bacteria unable to digest cellulose at low pH? *J. Dairy Sci.* 79(8), 1503–9. doi:10.3168/jds.S0022-0302(96)76510-4.

Sadet, S., Martin, C., Meunier, B. and Morgavi, D. P. 2007. PCR-DGGE analysis reveals a distinct diversity in the bacterial population attached to the rumen epithelium. *Animal* 1(7), 939–44. doi:10.1017/S1751731107000304.

Salvati, G. G. S., Morais Junior, N. N., Melo, A. C. S., Vilela, R. R., Cardoso, F. F., Aronovich, M., Pereira, R. A. and Pereira, M. N. 2015. Response of lactating cows to live yeast supplementation during summer. *J. Dairy Sci.* 98(6), 4062–73. doi:10.3168/jds.2014-9215.

Schiavone, M., Sieczkowski, N., Castex, M., Dague, E. and Marie François, J. 2015. Effects of the strain background and autolysis process on the composition and biophysical properties of the cell wall from two different industrial yeasts. *FEMS Yeast Res.* 15(2). doi:10.1093/femsyr/fou012.

Schiavone, M., Déjean, S., Sieczkowski, N., Castex, M., Dague, E. and François, J. M. 2017. Integration of biochemical, biophysical and transcriptomics data for investigating the structural and nanomechanical properties of the yeast cell wall. *Front. Microbiol.* 8, 1806. doi:10.3389/fmicb.2017.01806.

Schofield, B. J., Lachner, N., Le, O. T., McNeill, D. M., Dart, P., Ouwerkerk, D., Hugenholtz, P. and Klieve, A. V. 2018. Beneficial changes in rumen bacterial community profile in sheep and dairy calves as a result of feeding the probiotic *Bacillus amyloliquefaciens* H57. *J. Appl. Microbiol.* 124(3), 855–66. doi:10.1111/jam.13688.

Segura, A., Bertoni, M., Auffret, P., Klopp, C., Bouchez, O., Genthon, C., Durand, A., Bertin, Y. and Forano, E. 2018. Transcriptomic analysis reveals specific metabolic pathways of enterohemorrhagic Escherichia coli O157:H7 in bovine digestive contents. *BMC Genomics* 19(1), 766. doi:10.1186/s12864-018-5167-y.

Seymour, E. H., Jones, G. M. and McGilliard, M. L. 1988. Persistence of residues in milk following antibiotic treatment of dairy cattle. *J. Dairy Sci.* 71(8), 2292–6. doi:10.3168/jds.S0022-0302(88)79806-9.

Shabat, S. K. B., Sasson, G., Doron-Faigenboim, A., Durman, T., Yaacoby, S., Berg Miller, M. E., White, B. A., Shterzer, N. and Mizrahi, I. 2016. Specific microbiome-dependent mechanisms underlie the energy harvest efficiency of ruminants. *ISME J.* 10(12), 2958–72. doi:10.1038/ismej.2016.62.

Signorini, M. L., Soto, L. P., Zbrun, M. V., Sequeira, G. J., Rosmini, M. R. and Frizzo, L. S. 2012. Impact of probiotic administration on the health and fecal microbiota of young calves: A meta-analysis of randomized controlled trials of lactic acid bacteria. *Res. Vet. Sci.* 93(1), 250–8. doi:10.1016/j.rvsc.2011.05.001.

Silanikove, N. 2000. Effects of heat stress on the welfare of extensively managed domestic ruminants. *Livest. Prod. Sci.* 67(1–2), 1–18. doi:10.1016/S0301-6226(00)00162-7.

Silberberg, M., Chaucheyras-Durand, F., Commun, L., Mialon, M. M., Monteils, V., Mosoni, P., Morgavi, D. P. and Martin, C. 2013. Repeated acidosis challenges and live yeast supplementation shape rumen microbiota and fermentations and modulate inflammatory status in sheep. *Animal* 7(12), 1910–20. doi:10.1017/S1751731113001705.

Soberon, F. and Van Amburgh, M. E. 2013. Lactation Biology Symposium: the effect of nutrient intake from milk or milk replacer of preweaned dairy calves on lactation milk yield as adults: a meta-analysis of current data. *J. Anim. Sci.* 91(2), 706–12. doi:10.2527/jas.2012-5834.

Soto, L. P., Frizzo, L. S., Signorini, M. L., Zbrun, M. V., Lavari, L., Bertozzi, E., Sequeira, G. and Rosmini, M. 2015. Faecal culturable microbiota, growth and clinical parameters of calves supplemented with lactic acid bacteria and lactose prior and during experimental infection with *Salmonella* Dublin DSPV 595T. *Arch. Med. Vet.* 47(2), 237–44. doi:10.4067/S0301-732X2015000200017.

Soto, L. P., Astesana, D. M., Zbrun, M. V., Blajman, J. E., Salvetti, N. R., Berisvil, A. P., Rosmini, M. R., Signorini, M. L. and Frizzo, L. S. 2016. Probiotic effect on calves infected with *Salmonella* Dublin: haematological parameters and serum biochemical profile. *Benef. Microbes* 7(1), 23–33. doi:10.3920/BM2014.0176.

Sousa, D. O., Oliveira, C. A., Velasquez, A. V., Souza, J. M., Chevaux, E., Mari, L. J. and Silva, L. F. P. 2018. Live yeast supplementation improves rumen fibre degradation in cattle grazing tropical pastures throughout the year. *Anim. Feed Sci. Technol.* 236, 149–58. doi:10.1016/j.anifeedsci.2017.12.015.

Souza, R. F. S., Rault, L., Seyffert, N., Azevedo, V., Le Loir, Y. and Even, S. 2018. *Lactobacillus casei* BL23 modulates the innate immune response in *Staphylococcus aureus*-stimulated bovine mammary epithelial cells. *Benef. Microbes* 9(6), 985–95. doi:10.3920/BM2018.0010.

Steele, M. A., Kroom, J., Kahler, M., AlZahal, O., Hook, S. E., Plaizier, K. and McBride, B. W. 2011. Bovine rumen epithelium undergoes rapid structural adaptations during grain-induced subacute ruminal acidosis. *Am. J. Physiol. Regul. Integr. Comp. Physiol.* 300(6), R1515–23. doi:10.1152/ajpregu.00120.2010.

Steele, M. A., Penner, G. B., Chaucheyras-Durand, F. and Guan, L. L. 2016. Development and physiology of the rumen and the lower gut: targets for improving gut health. *J. Dairy Sci.* 99(6), 4955–66. doi:10.3168/jds.2015-10351.

Stelwagen, K., Carpenter, E., Haigh, B., Hodgkinson, A. and Wheeler, T. T. 2009. Immune components of bovine colostrum and milk. *J. Anim. Sci.* 87(13 Suppl.), 3–9. doi:10.2527/jas.2008-1377.

Stier, H. and Bischoff, S. C. 2016. Influence of *Saccharomyces boulardii* CNCM I-745 on the gut-associated immune system. *Clin. Expl. Gastroenterol.* 9, 269–79. doi:10.2147/CEG.S111003.

Sullivan, M. L. and Bradford, B. J. 2011. Viable cell yield from active dry yeast products and effects of storage temperature and diluent on yeast cell viability. *J. Dairy Sci.* 94(1), 526–31. doi:10.3168/jds.2010-3553.

Sun, X., Yuan, X., Chen, L., Wang, T., Wang, Z., Sun, G., Li, X., Li, X. and Liu, G. 2017. Histamine induces bovine rumen epithelial cell inflammatory response via NF-κB pathway. *Cell. Physiol. Biochem.* 42(3), 1109–19. doi:10.1159/000478765.

Tabe, E. S., Oloya, J., Doetkott, D. K., Bauer, M. L., Gibbs, P. S. and Khaitsa, M. L. 2008. Comparative effect of direct-fed microbials on fecal shedding of *Escherichia coli* O157:H7 and *Salmonella* in naturally infected feedlot cattle. *J. Food Prot.* 71(3), 539–44. doi:10.4315/0362-028X-71.3.539.

Terré, M., Maynou, G., Bach, A. and Gauthier, M. 2015. Effect of *Saccharomyces cerevisiae* CNCM I-1077 supplementation on performance and rumen microbiota of dairy calves. *Prof. Anim. Sci.* 31(2), 153–8. doi:10.15232/pas.2014-01384.

Tkalcic, S., Zhao, T., Harmon, B. G., Doyle, M. P., Brown, C. A. and Zhao, P. 2003. Fecal shedding of enterohemorrhagic *Escherichia coli* in weaned calves following treatment with probiotic *Escherichia coli*. *J. Food Prot.* 66(7), 1184–9. doi:10.4315/0362-028x-66.7.1184.

Trevisi, E. and Minuti, A. 2018. Assessment of the innate immune response in the periparturient cow. *Res. Vet. Sci.* 116, 47–54. doi:10.1016/j.rvsc.2017.12.001.

Trevisi, E., Riva, F., Filipe, J. F. S., Massara, M., Minuti, A., Bani, P. and Amadori, M. 2018. Innate immune responses to metabolic stress can be detected in rumen fluids. *Res. Vet. Sci.* 117, 65–73. doi:10.1016/j.rvsc.2017.11.008.

USDA. 2014a. Dairy 2014: health and management practices on U.S dairy operations.

USDA 2014b. Dairy 2014: dairy cattle management practices in the United States.

Uyeno, Y., Sekiguchi, Y., Tajima, K., Takenaka, A., Kurihara, M. and Kamagata, Y. 2010. An rRNA-based analysis for evaluating the effect of heat stress on the rumen microbial composition of Holstein heifers. *Anaerobe* 16(1), 27–33. doi:10.1016/j.anaerobe.2009.04.006.

Uyeno, Y., Akiyama, K., Hasunuma, T., Yamamoto, H., Yokokawa, H., Yamaguchi, T., Kawashima, K., Itoh, M., Kushibiki, S. and Hirako, M. 2017. Effects of supplementing an active dry yeast product on rumen microbial community composition and on subsequent rumen fermentation of lactating cows in the mid-to-late lactation period. *Anim. Sci. J.* 88(1), 119–24. doi:10.1111/asj.12612.

Villageliú, D. and Lyte, M. 2018. Dopamine production in *Enterococcus faecium*: a microbial endocrinology-based mechanism for the selection of probiotics based on neurochemical-producing potential. *PLoS ONE* 13(11), e0207038. doi:10.1371/journal.pone.0207038.

Villena, J., Aso, H., Rutten, V. P. M. G., Takahashi, H., van Eden, W. and Kitazawa, H. 2018. Immunobiotics for the bovine host: their interaction with intestinal epithelial cells and their effect on antiviral immunity. *Front. Immunol.* 9, 326. doi:10.3389/fimmu.2018.00326.

Villot, C., Meunier, B., Bodin, J., Martin, C. and Silberberg, M. 2018. Relative reticulo-rumen pH indicators for subacute ruminal acidosis detection in dairy cows. *Animal* 12(3), 481-90. doi:10.1017/S1751731117001677.

Vipham, J. L., Loneragan, G. H., Guillen, L. M., Brooks, J. C., Johnson, B. J., Pond, A., Pond, N. and Brashears, M. M. 2015. Reduced burden of *Salmonella enterica* in bovine subiliac lymph nodes associated with administration of a direct-fed microbial. *Zoonoses Public Health* 62(8), 599-608. doi:10.1111/zph.12187.

von Buenau, R., Jaekel, L., Schubotz, E., Schwarz, S., Stroff, T. and Krueger, M. 2005. *Escherichia coli* strain Nissle 1917: significant reduction of neonatal calf diarrhea. *J. Dairy Sci.* 88(1), 317-23. doi:10.3168/jds.S0022-0302(05)72690-4.

Vyas, D., McGeough, E. J., Mohammed, R., McGinn, S. M., McAllister, T. A. and Beauchemin, K. A. 2014. Effects of Propionibacterium strains on ruminal fermentation, nutrient digestibility and methane emissions in beef cattle fed a corn grain finishing diet. *Animal* 8(11), 1807-15. doi:10.1017/S1751731114001657.

Wallis, J. K., Krömker, V. and Paduch, J. H. 2018. Biofilm formation and adhesion to bovine udder epithelium of potentially probiotic lactic acid bacteria. *AIMS Microbiol.* 4(2), 209-24. doi:10.3934/microbiol.2018.2.209.

Wallis, J. K., Krömker, V. and Paduch, J. H. 2019. Biofilm challenge: lactic acid bacteria isolated from bovine udders versus staphylococci. *Foods* 8(2). doi:10.3390/foods8020079.

Wang, D. S., Zhang, R. Y., Zhu, W. Y. and Mao, S. Y. 2013. Effects of subacute ruminal acidosis challenges on fermentation and biogenic amines in the rumen of dairy cows. *Livest. Sci.* 155(2-3), 262-72. doi:10.1016/j.livsci.2013.05.026.

Wehnes, C. A., Novak, K. N., Patskevich, V., Shields, D. R., Coalson, J. A., Smith, A. H., Davis, M. E. and Rehberger, T. G. 2009. Benefits of supplementation of an electrolyte scour treatment with a Bacillus-based direct fed microbial for calves. *Probiotics Antimicrob. Proteins* 1(1), 36-44. doi:10.1007/s12602-008-9004-5.

Wisener, L. V., Sargeant, J. M., O'Connor, A. M., Faires, M. C. and Glass-Kaastra, S. K. 2015. The use of direct-fed microbials to reduce shedding of *Escherichia coli* O157 in beef cattle: a systematic review and meta-analysis. *Zoonoses Public Health* 62(2), 75-89. doi:10.1111/zph.12112.

Yáñez-Ruiz, D. R., Abecia, L. and Newbold, C. J. 2015. Manipulating rumen microbiome and fermentation through interventions during early life: a review. *Front. Microbiol.* 6, 1133. doi:10.3389/fmicb.2015.01133.

Yeoman, C. J., Ishaq, S. L., Bichi, E., Olivo, S. K., Lowe, J. and Aldridge, B. M. 2018. Biogeographical differences in the influence of maternal microbial sources on the early successional development of the bovine neonatal gastrointestinal tract. *Sci. Rep.* 8(1), 3197. doi:10.1038/s41598-018-21440-8.

Yohe, T. T., Enger, B. D., Wang, L., Tucker, H. L. M., Ceh, C. A., Parsons, C. L. M., Yu, Z. and Daniels, K. M. 2018. Short communication: does early-life administration of a *Megasphaera elsdenii* probiotic affect long-term establishment of the organism in the rumen and alter rumen metabolism in the dairy calf? *J. Dairy Sci.* 101(2), 1747-51. doi:10.3168/jds.2017-12551.

Yuan, K., Mendonça, L. G. D., Hulbert, L. E., Mamedova, L. K., Muckey, M. B., Shen, Y., Elrod, C. C. and Bradford, B. J. 2015. Yeast product supplementation modulated humoral and mucosal immunity and uterine inflammatory signals in transition dairy cows. *J. Dairy Sci.* 98(5), 3236-46. doi:10.3168/jds.2014-8469.

Zebeli, Q., Terrill, S. J., Mazzolari, A., Dunn, S. M., Yang, W. Z. and Ametaj, B. N. 2012. Intraruminal administration of *Megasphaera elsdenii* modulated rumen fermentation profile in mid-lactation dairy cows. *J. Dairy Res.* 79(1), 16–25. doi:10.1017/S0022029911000707.

Zebeli, Q., Ghareeb, K., Humer, E., Metzler-Zebeli, B. U. and Besenfelder, U. 2015. Nutrition, rumen health and inflammation in the transition period and their role on overall health and fertility in dairy cows. *Res. Vet. Sci.* 103, 126–36. doi:10.1016/j.rvsc.2015.09.020.

Zhao, T., Tkalcic, S., Doyle, M. P., Harmon, B. G., Brown, C. A. and Zhao, P. 2003. Pathogenicity of enterohemorrhagic *Escherichia coli* in neonatal calves and evaluation of fecal shedding by treatment with probiotic *Escherichia coli. J. Food Prot.* 66(6), 924–30. doi:10.4315/0362-028x-66.6.924.

Zhu, Z., Noel, S. J., Difford, G. F., Al-Soud, W. A., Brejnrod, A., Sørensen, S. J., Lassen, J., Løvendahl, P. and Højberg, O. 2017. Community structure of the metabolically active rumen bacterial and archaeal communities of dairy cows over the transition period. *PLoS ONE* 12(11), e0187858. doi:10.1371/journal.pone.0187858.

Zhu, Z., Kristensen, L., Difford, G. F., Poulsen, M., Noel, S. J., Abu Al-Soud, W., Sørensen, S. J., Lassen, J., Løvendahl, P. and Højberg, O. 2018. Changes in rumen bacterial and archaeal communities over the transition period in primiparous Holstein dairy cows. *J. Dairy Sci.* 101(11), 9847–62. doi:10.3168/jds.2017-14366.

www.ingramcontent.com/pod-product-compliance
Lightning Source LLC
Chambersburg PA
CBHW050534270326
41926CB00015B/3214